Engineering a New Life
FROM COMPLACENCY TO CLARITY

Steve Gurklys

Copyright © 2018 Steve Gurklys

All rights reserved. This book or any portion thereof may not be reproduced or used in any manner whatsoever without the express written permission of the publisher except for the use of brief quotations in a book review.

123 Thinking Press
contact@123thinkingpress.com
www.123thinkingpress.com

ISBN-13: 978-1-7325439-0-4

Table of Contents

- Table of Contents 3
- List of Figures 9
- Acknowledgements 11
- Prologue 13
- Getting Ready 15
  - How to use this book 15
  - EXERCISE – Get your tools! 16
- How it Started 17
  - Disclosure 18
  - My Credentials 18
  - An Engineer's Perspective 18
  - Decide Now! 20
- Where am I? 21
  - Me, Myself and I 21
  - What is my purpose? 23
  - Where do you start? 23
  - EXERCISE – You just won the lottery! (10 min) 25
  - What do you want? 31
  - EXERCISE – Life Goals (15 min) 32
  - A la Carte Goals 33
  - EXERCISE – Life Goal Review (30 min) 34
  - Reality Check 35
  - Checking for Balance 36
  - EXERCISE – Category Coloring (15 min) 37
  - Supporting Your Goals 38
- We are Robots 39
  - Carbon vs. Silicon 39
  - We Are Just a Story 40
  - Why did I just do that?!! 40
  - EXERCISE – Count the F's (2 min) 42
  - Observe and Know Thyself 42
  - EXERCISE – Reflection (30 seconds) 43
  - EXERCISE – Reflection Now (15-30 min) 44
  - Life Outside the Matrix 44
  - Opportunities are everywhere! 45
  - Instant Gratification 45
  - Programming 45
  - Human Programming 48
  - Making Choices 51
  - EXERCISE – Power of Affirmation (30 min) 53
  - Software Debugging 54
  - Nature of Human Interaction 55
- Obstacles 57
  - Don't Have Time 57
  - EXERCISE – Time Analysis (30 min) 57
  - Life on Autopilot 58

- Limited by Commitments 58
- Lack of Confidence 58
- Self-Sabotage 59
- Addictions 60
- Co-Dependent Relationship 60
- Quest for Perfection 61
- Dysfunctional Family 62
- EXERCISE – Determining Dysfunction (10 min) 62
- Anything Else? 63
- EXERCISE – Other Obstacles (20 min) 64
- Second Opinion 64
- What do you fear? 64
- EXERCISE – Monster Exorcism (30 min) 67
- Uncertainty 68

Changing Behaviors 69
- 90 Second Rule 69
- Accountability 71
- Conscious Living 71
- Who You Need to Change 71
- Judging Others 71
- Small Increments 72
- Dealing with Numbing & Making Time 72
- Dealing with Commitments 72
- Getting Confidence 73
- EXERCISE - Confidence (15 min) 74
- Dealing with Self-Sabotage 75
- Scenario 1 – Absent Adam 77
- Scenario 2 – Awkward Adam 78
- Scenario 3 – Active Adam 78
- EXERCISE – Drawing (15 min) 79
- Dealing with Addictions 79
- Dealing with Controlling Relationship 80
- Negative Thoughts 80
- Dealing with Perfection 81
- Dysfunctional Family of Origin 82

A New Direction 85
- Challenging Assumptions 85
- EXERCISE – Checking Assumptions (30 min) 90
- Scarcity and Abundance 91
- Credis Quod Habes 92
- You can do anything 93
- Rainbows and Unicorns – Not! 94
- Priorities 94
- Building an Amazing Life 96
- Affirmations 97
- Spin It Around 98
- Brainstorming 98
- EXERCISE – Brainstorming (60 min) 100
- Give it a Rest! 100

Scientific Process 101
- Inside the Mind of a Scientist 101

The Scientific Method	*101*
Occam's Razor	*105*
Personal Progress Method	*106*
EXERCISE – Plan for #1 Life Goal (60 min)	*108*

Say Hello to your Right Brain 111
EXERCISE – Soul Smiling Quotient (15 min)	*111*
Geckos	*117*
Get in touch with the Universe	*117*
EXERCISE - Visiting your Right Brain (45 min)	*119*
How it works	*119*
Meditation	*120*
EXERCISE – First Meditation (15 min)	*121*
Learn to Play	*122*
Power of Music	*122*
Delight of Dancing	*122*
EXERCISE – Art Goal Review (15 min)	*123*

Calories In, Calories Out 125
Enjoy the Moment	*125*
Designed to Move	*125*
It's not too late	*126*
Start Today!	*127*
Diet	*128*
Magic Calorie Equation	*129*
Calories Eaten – Exercise = Weight Gain (Loss)	*129*
EXERCISE – Diet/Exercise Choice (1 min)	*130*
Nutrition Facts Label	*130*
Burning the Calories	*131*
Calorie Counting - Not!	*131*
Costa Rica	*133*
Which one is Better	*134*
Macro Nutrients	*135*
Natural Options	*137*
Sugar	*138*
Variety	*138*
EXERCISE – Calorie Cost (60 min)	*138*
The Sweaty Kind of Exercise	*139*
EXERCISE – Health Goal Review (15 min)	*139*
Living the Moment	*142*
Shark Story	*142*
The Key to the Toolbox	*143*
Gecko Animal Guide	*144*
Back to the Beach	*145*
Wake up to the life around you	*146*
Innocence	*147*
EXERCISE – Meeting People (For Adults Only)	*148*
Do Something Spiritual	*148*
EXERCISE – Spiritual Goal Review (15 min)	*148*

Human Connection 149
Our Purpose on the Planet	*149*
Why do people cave-in?	*150*
Getting Unreasonable Happiness	*151*

Closed Book 151
Confidence 152
Communication 153
EXERCISE – Communication Recall (20 min) 154
Connection 154
Making Friends 155
Making and Keeping Friends 157
EXERCISE – Get to know someone (60 min) 158
Alienating Friends 158
Kids 158
EXERCISE – People Goal Review (15 min) 160

Romance **161**
The Courtroom 161
What is Love Anyways 161
A New Start 162
Blissful Biology 162
Soul Mate 163
Seven Year Itch 164
Right and Right Away 164
You are in charge of your feelings 166
Online Dating 166
Downtown Disney 168
EXERCISE – Romance Goal Review (30 min) 169

Home **171**
Feng shui 171
EXERCISE – De-Clutter (60 min) 174
Finding Your Home 174
EXERCISE – Home Specifications (60 min) 176
EXERCISE – Home Goal Review (30 min) 177

Money, Job, Career **179**
Simple Concepts that helped 179
Financial Health Equation 179
Income – Costs = Financial Health 179
Time – Money Equation 180
Time you give up = Money saved 181
Money Spent = Time you gain 181
A Fresh Start 182
EXERCISE – Not Happy with my Job (30 min) 183
Memos and Meetings 184
Follow your Dream 184
Lessons Learned 185
Making Choices that are right for you 186
EXERCISE – Analyze your Career (60 min) 187
Be your Own Boss 188
8 Essential Business Plan Elements 188
EXERCISE – Career Goal Review (30 min) 189

Closing Thoughts **191**
Convergence 191
"To Do" Lists 191
Less Thinking, more Feeling 193
Regular Check Ups 193

EXERCISE – Display Your Goals (30 min)	*194*
A Passion for Everything	*197*
Feedback	*198*
Conclusion	*198*
Answers to the Exercises	**199**
You just won the lottery!	*199*
Life Goals	*199*
Life Goal Review	*199*
Category Coloring	*199*
Counting the F's	*201*
Reflection	*201*
Reflection Now	*201*
Power of Choice Affirmation	*201*
Time Analysis	*201*
Other Obstacles	*204*
Monster Exorcism	*204*
Confidence	*204*
Drawing	*204*
Checking Assumptions	*204*
Brainstorming	*204*
Plan for #1 Life Goal	*205*
Soul Smiling Quotient	*205*
Visiting your Right Brain	*205*
First Meditation	*205*
Art Goal Review	*205*
Diet/Exercise Choice	*205*
Calorie Cost	*205*
Health Goal Review	*206*
Meeting People	*206*
Spiritual Goal Review	*206*
Communication Recall	*206*
Golden Rectangle	*206*
Get to know someone	*209*
People Goal Review	*209*
Romance Goal Review	*209*
De-Clutter	*210*
Home Spec	*210*
Home Goal Review	*210*
Not happy with my job	*210*
Analyze your Career	*210*
Career Goal Review	*211*
Display your Goals	*211*
References	**213**

Engineering a New Life:

List of Figures

Engineering a New Life Roadmap	15
What You Can Change	24
Seven-Segment Life Wheel	33
Author's Colored Life Wheel	36
Stop Driving on Autopilot	39
Don't Let Other People Push Your Buttons	41
Water Sprinkler Programming Example	46
Water Sprinkler Flow Diagram	47
Mary Sue Campbell Brain Loop	49
Typical Daily Routine Flow Chart	51
Time Analysis Chart	58
Let your inspiration take you on a ride!	60
Some Fear Manifestations	65
The 90 Second Rule to Change your Behavior	70
So many doors are available to you – pick one!	77
ACA Laundry List Summary	83
Let there be water!	85
Periodic Table of the Elements	86
Peering Inside the Atom	86
Peering Inside the Atom - Setup	87
Peering Inside the Atom - Experiment	87
Peering Inside the Atom - Conclusion	88
An abundant universe raises all hopes	92
Life is a string of moments	96
A Special Sunset Affirmation	97
Scientific Method Flowchart	101
Falling Objects Flowchart – Version 1	104
Falling Objects Flowchart – Version 2	104
Ball Drop Equation and Table	105
Personal Progress Method Flowchart	106
Top Life Goal Plan	109
Left and Right Brain Functions	118
Drawings by Left (analytical) and Right (creative) Brains	120
Typical Adult Male Life – Then and Now	126
Nutritional Facts for Large Bag of Potato Chips	131
Nutritional Facts Labels	136
Macro Nutrients for 4 Ounce Servings	136
Spiritual Awakening Comes from Within	141
Inspiration is all around you	145
Everyone has a story	149
Windows to the soul	154
Golden Rule for a Golden Connection	155
Bagua to enhance your home and goals	172
Dalle de Verre Dread to Delight	173
Life Goals Summary Sheet 1	195
Life Goals Summary Sheet 2	196
Two Golden Rectangles from 2 Solutions	209

Acknowledgements

Many people have provided support, inspiration and kindness on my journey. This book would not have been possible without each and every one of them – thank you! ACA and everyone who attended the "Anything Goes" chapter provided a nurturing environment where many of the concepts in this book were developed and tested. I am most appreciative to Pilar for forming this group, always ready to lend a supporting ear and warm smile that lit a torch of hope for me and many others.

I am so grateful to David my fellow traveler without whom the road would have been far rockier and fraught with many more detours. He helped refine many of the ideas in the book, most importantly the subtitle "From complacency to clarity" that he brilliantly suggested which provided, well … clarity.

Dr. Paul McCandless and Dr. Charles Coull, my therapists, were instrumental in my life awakening. Their gentle candor and wisdom, gave me just enough insight to get through the week and a glimpse to future possibilities. They were far more helpful than I could have possibly imagined.

A special thank you to my beautiful and wonderful Serrah, my partner in dance and a whole lot more who makes the world a little brighter for all those whom she touches – especially me!

Prologue

After a three-month explosion of ideas and words that cascaded into the manuscript for this book, I felt an immense calm and contentment. Although certain to publish it one day, I knew it was not a priority. Getting all my thoughts on the page enabled me to crystalize my own "Life Wheel", complete with areas that needed some attention: Home and Romance. And now it was time to heed my own advice and engineer a complete well-rounded life for myself.

It's now August 2017 and I have not touched the manuscript in over two years. I purchased a nice house, made some modifications including a 32-foot garage bench and ping pong table in the living room to make it mine. So check the "Home" section of the Life Wheel.

As for the "Romance" life category … how can I explain it … have you ever watched a romantic movie and smiled when the lovers met? A memorable example for me is Tom Hanks and Meg Ryan touching hands for the first time atop the Empire State Building during "Sleepless in Seattle". Have you lingered in the theater during the credits because you're not quite ready to get back to reality? Or dreamt of the iconic "Gone with the Wind" scene where Rhett Butler carries Scarlett in her elegant crimson flowing gown up the alabaster stairs to the bedroom and wished to have such a person in your life?

I feel that way every single time my beautiful goddess appears, every time I hear her sexy voice, every kiss. My life has become a romantic movie. And I tell her how lucky and grateful I feel – every day. You can read about our storybook meeting in the coming pages. I thought about writing a screenplay of our romance, but frankly, no actress could possibly do justice to her beauty, charm and grace.

It's now time to share this book with you.

Steve Gurklys

Getting Ready

How to use this book

Please don't feel compelled to read every word in this book. Skip the chapters that are not useful to you and take my suggestions as they are intended – just suggestions; figure out what works best for you. Treat the book as a tool to guide you on your journey. Here's a map of the chapters in this book to navigate to what you need.

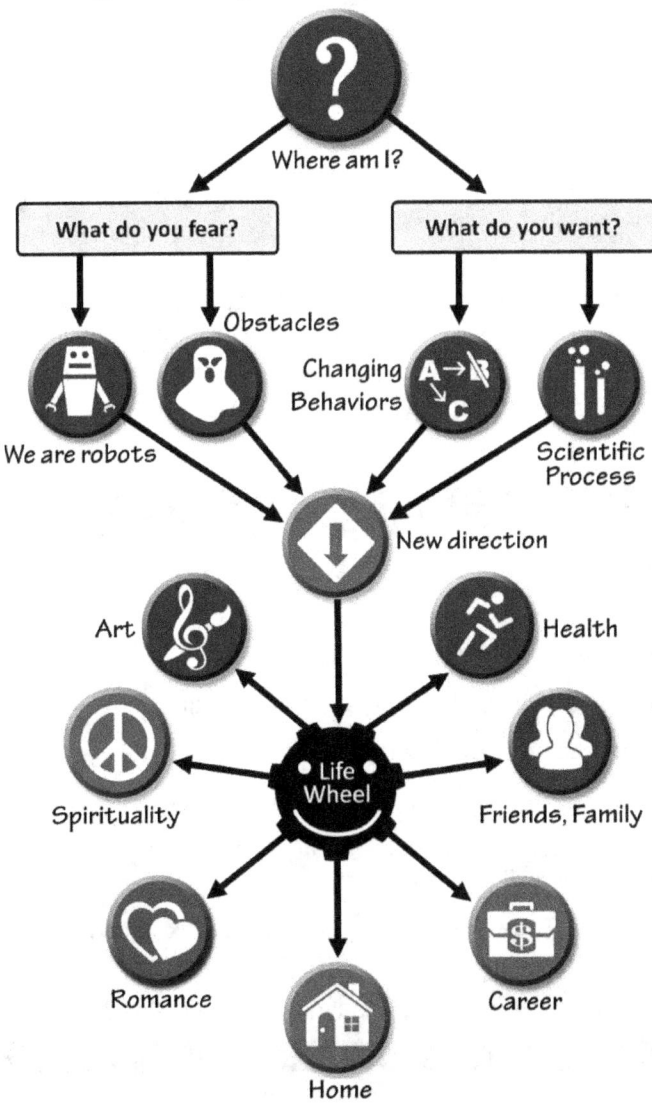

The questions in this book are for the most part not rhetorical; they are intended for you to answer. I would suggest that you answer them out loud. You may be surprised at what you hear; then write down your responses before they slip through your fingers. There are exercises in this book that will help you with this process. Use pencil for most of them since you may change some of your answers as you progress through the book.

Doodle, dog ear, cross out, make margin notes, highlight sections you like, cross out sections you don't, add yellow self-stick notes - whatever inspiration strikes you. Your marks will help personalize the experience and you will get more out of the material. When I'm reading a book and find a really good paragraph I start my own index inside the front cover with a phrase and page number to find it easily next time I pick up the book. Life is messy – so should be the workbook for your life goals.

As I've been telling students for over 25 years – you have to do the exercises. Learning studies performed in the 1960s revealed that we retain only 10% of what we read[1]. When you hear it out loud your comprehension doubles to 20%. But when you write it down and perform exercises your understanding soars to 70%! In my face to face training sessions, primarily for engineers, it is so gratifying when a student says: *"Oh, so that's how it works ... I get it now!"* My hope is that you have the same reaction when you complete the exercises in this book.

EXERCISE – Get your tools!

Here are the supplies you will need. Assemble them now so you're ready for the exercises.

- Pencil, because you will be writing
- Pencil Sharpener, because you will be writing a lot
- Eraser, because some things in your plan will change
- Pen, because some things will always be true
- Set of colored crayons, markers or pencil crayons, because we will be making art
- Scissors & Tape, because no self-help book is complete without cut & paste
- Timer, because we want to have a better life ASAP!

How it Started

I cracked open a fortune cookie that proclaimed *"Good ideas will spring forth naturally from your mind in the coming week"*. I was beaming when I read it and immediately taped it to the bottom of my monitor. I had written down a note about maybe writing about my experiences that have completely changed my life at some point. But shortly after reading that fortune cookie – Bam! I started writing down thoughts, phrases and full paragraphs – it was cathartic and the words just kept coming. The lessons from my tumultuous 19 months started pouring out.

The next day I walked to my local coffee shop and wrote for over an hour. I then went to the local farmer's market on Del Mar and was compelled to stop, sit on a bench and keep writing on a crumpled sheet of paper overwhelming every white space with words. Later I was driving on some errands and absolutely had to pull over to write for another 30 minutes. After three weeks of such bursts of insights, I had written 20,000 words (approximately 70 pages). I reached that milestone while landing in Denver during a summer storm. A few days later while writing these words in the Blue Café and chatting with Bri, a budding Psychologist, I was completely committed to publishing my experiences.

Weeks later during a Baltimore training class I went to the local mall to a Chinese restaurant for lunch. After enjoying my cashew nut beef, I cracked open another fortune cookie and thanked the universe out loud; the message read: *"Your present plans will succeed"* – creating this book!

Writing the manuscript was really an exercise for me to document my metamorphosis so I would not forget it. (Same training study referenced previously showed that if you teach someone your retention goes to 90%). Each of us is unique and I don't pretend to know you, your history or what you want out of life. My goal in writing this book is to simply share with you what I did to turn my life around and show you some tools that may help you too. I believe this book is unlike any other you have ever read; it is part biography, part instructional self-help, part insight into the scientific mind, sprinkled with unbridled optimism.

Disclosure

I'm not a doctor, therapist, nutritionist or financial planner but nevertheless I have collected a bag of tricks that have worked exceedingly well for me and I want to share them with you. These assorted thoughts, techniques, and exercises along with help from many people have proven to be highly effective in turning my life completely around. I feel really good about myself and the world around me, every day embracing life with gusto and eager to get out of bed! At times this is a challenge because I also don't want to stop playing the piano until late at night.

I don't profess to be an expert in anything except me – and that's a relatively new area of knowledge. I hope that my stories, examples, questions and exercises help you explore your situation and give you a framework to start arranging your home, career, friends/family, health, art, spirituality and romance for a happier life.

My Credentials

I have chalked up over 50 years on the planet so I definitely have some experiences to share. A 27-year marriage indicates commitment; a divorce of that same marriage is my claim to failure. A rather spectacular failure actually – but we will get to that a little later. I was raised in a dysfunctional family that left deep scars affecting my adult life. Then I had a spiritual awakening (very cool) and finally started putting my life together. In some ways, I am now truly living for the first time.

As an engineer, instructor and training developer with more than 25 years' experience, I can convey complex ideas in a simple and logical progression; this book is easy to understand, follow and will guide you – if you're ready. It's packed with graphics, exercises and fun because interactivity is essential to making learning actually work.

An Engineer's Perspective

How many engineers does it take to change a light bulb? None. The engineering team will design, test and train a robot to do it because that would be far more efficient.

Most of us engineers wear the nerd moniker proudly and are completely oblivious to daily interactions of the real world; that place is rife with feelings and emotions – really uncomfortable. And simple things can be downright baffling. For example, how do you correctly sort colors from whites for the laundry? As soon as I find a piece of clothing that has a single speck of white I'm immediately tempted to pop it in the "whites" bin. But when I do, there's not much left for the "colors" bin. If on the other hand I consider a sort based on "brights" vs. "darks", then I need to know the threshold brightness coefficient so that I can apportion the items accordingly. I have stared at a shirt for 15 seconds wondering about the correct bin assignment for this item. Get the idea? I just did my best guess and gave it up to my higher power to ensure that it all works out ... in the wash.

So why should you be interested in an engineer's perspective on life? Engineers spend much of their time solving theoretical, esoteric problems. In part, I believe this ability stems from a perspective of no pre-conceptions so any scenario is possible – that's one of the reasons engineers generally love Science Fiction. This is also why we struggle in social situations with behavior rules that everyone seems to know inherently – we largely don't get it. Or if we do, it's been a painful process that has been explained to us. For example, my ex patiently stated many years ago that in the presence of a house guest it is not okay to start debugging a circuit board if you don't find the conversation interesting. (Yes, seriously I did that). But we do come in handy when there's an unusual situation that social norms don't cover because we can access the relevant parameters and figure out the best course of action.

My life imploded in a heap of ashes when my marriage ended and there was no other choice other than to start again from the floor – literally; I slept on the floor in the hallway for over 6 months. Engineers solve problems. And I was faced with the biggest problem ever: *How do I engineer a life that is satisfying, noble and glorious?* Engineers are also good at documenting so you can benefit from my experience. This book examines the big picture of creating such a new life and provides a blueprint or roadmap to help you do the same with yours. I believe the exercises in this book will assist you to figure out what is going on in your life and re-assemble it into one you love. *"Without clarity and honesty we do not progress"*[2].

Because I never learned how to connect with people before, I was oh so eager to do it now. If you need assistance in this area, you will see how I did it from my humble beginnings. Believe it or not, we all yearn for connection overtly, secretly or unknowingly. Think about it – when was the last happy time you had joyful, gleeful, giggling and grinning ear to ear like you used to do when you were five? It almost certainly involved someone you care about and sharing a special moment with them. This connection is second nature to many people – but not for me. So I had to learn this and I believe that I can help those of you who also find this perplexing.

As an engineer, the solution now seems so darn simple. No, I'm not a genius, I just can see clearly from the blank slate I was gifted – a life reset. It occurred to me one day while working out in the gym, we can begin to achieve peace and serenity if we answer two questions honestly and thoroughly:

> **What do you want?**
>
> **What do you fear?**

Everything else is implementation details – and engineers have a reputation for dealing with the details and solving problems. Central to our engineer identity is coming up with solutions that work.

Decide Now!

This book is a practical guide to creating the life you want and deserve based on my personal experiences. This is the journey I want to take you on. Want to come along? It's your life – engineer yourself a better one. Be bold and take the first step - *right now!*

I commit to the process by dedicating _____ minutes every week to engineer my new life, one that I want and deserve.

_____ _____

(Sign here)   Date

Fellow traveler, I wish you great success on your journey!

Where am I?

For me that was one of the first questions that propelled me on my current path. But it wasn't like being whisked away by a tornado and ending up in a land of Munchkins and flying monkeys. Rather, it was realizing that slowly, over many years I eased into complacency about my life. I seemed to be on autopilot just going through the motions with negligibly little engagement with the world and a constant feeling of profound sadness – on the outside forlornly looking in. It was working – I had a job, a long-term marriage, reasonable health, living in southern California with a view of the ocean and 5 parrots. On the surface it seemed pretty good – but on closer examination it really wasn't.

My complacent life was a low-level but constant dysfunction. This steady erosion carved a channel in my soul like the Grand Canyon, only this one was forged in 40 years rather than 6 million. I felt like a stranger in my own life without a clue how I came to be here or how to get out. Heck, it felt that this was my lot in life and there were no options.

A friend referred me to an exquisitely creative website called "The Dictionary of Obscure Sorrows"[3]. It is a compilation of exquisite new words that define situations and feelings commonly experienced but are not effectively described in conventional dictionaries. I was astounded to find *Nodus Tollens* as it described exactly how I was feeling – suddenly discovering that you don't know where you are or how you arrived there.

It appeared that I was not the only one on the planet that felt so lost, confused and unconnected. That was a start.

Me, Myself and I

We are not the only creatures that are self-aware, that is to say an entity separate from the world around us. The mark test or mirror test[4] is a clever experiment devised by psychologist Gordon Gallup Jr. in 1970 to figure out if a chimpanzee was self-aware. After mildly sedating an unsuspecting primate, he dappled a red dot on its forehead and ensured a mirror was located nearby. If the chimpanzee reached out to the mirror to examine the red adornment, it would indicate that he thinks the image is another animal. But if he reaches to his own forehead to investigate the red blob that would clearly indicate he knows that he's looking at himself.

Upon awakening and locating the mirror, the subject peered at the unusual red blob (probably the way I check out my face in the morning to see if I need to shave) and nonchalantly reached to wipe his brow — he is self-aware!

Although debate continues as to when humans became self-aware, the answer seems to be converging around 60,000 years ago[5]. Beyond self-awareness, introspection consciousness is the ability to ponder our thoughts and feelings. Monkeys can visualize ideas but they do not appear to be able to manipulate them[6]. Only humans can manipulate ideas in our head — our imagination. For example, imagine a purple elephant that is 50 feet tall and has feathers. Even though no such creature has ever existed, we can imagine it, draw it and make a blockbuster movie about it. This is the power we have to manipulate ideas in our heads.

The brain we wake up to in the morning is not the brain that we put on our pillow at night. Our brains are dynamic - there are small connection changes as a result of experiencing and processing stimulus from the world around us. For example, you may wake up one morning and decide that it would be cool to learn how to read body language and undertake to do so. I highly recommend "The Definitive Book of Body Language"[7]. So you read a book or two, spend 15 minutes a day watching people's posture, hand positions and note their interactions with others. Within a few weeks you naturally pick up these not so subtle clues and have better people exchanges because you now know those that are more receptive. Your brain has now been rewired with this new skill.

According to neurologist Paul Brocks: "We're a car crash or slip away from being a completely different person"[8]. He believes that any major head injury can alter our brain connections which define our memories, associations and behaviors. You probably have heard of patients with memory loss due to severe head trauma.

There was a most intriguing case of a 46 year old woman who experienced a brain aneurism that rendered her unconscious for three months[9]. After extensive care and seven months of rehabilitation to relearn many basic skills such as eating and walking, she was released from the hospital. Her daughter recounts how she felt that her mother had died that first night she arrived at the hospital. Her mother regained all her memories of her life, her daughter and how she used to be, except that a completely new personality slowly emerged. Where before she was a sedate, socially proper woman, she now displayed a tattoo, loved sex and sported Groucho Marx glasses while singing *"Hey Good Looking,* what you got cooking?" at airport gate. So if our brains are malleable and change in the course of our daily interactions with the world ... perhaps I can direct these changes in a positive way in my own head ... perhaps ...

What is my purpose?

Not having an adequate answer to this question kept me firmly in the morass of apathy and complacency. It usually was followed by additional questions such as "why bother?" and "what's the point?"

Carl Sagan had a simple answer to the grandiose soul searching question: *"What is my purpose?"* He said simply and with a little annoyance: *"Do something worthwhile!"*[10] In my opinion it's good advice because it is simple, clear and can be acted on. My interpretation is not to worry about leading the perfect life and discovering the one and only path that is the chosen one for you. Rather, just pick one and get on with it! Choose a goal that excites you, that is positive and touches or helps others. You can then start living and moving forward – right now. And if it turns out it's not what you really want, no problem – pick another one! F. Scott Fitzgerald said it even better: *"I hope you live a life you are proud of; if you find you are not, I hope you find the strength to start over all over again."* Two years ago someone gave me this quote and I am grateful to them because now, today, I can whole-heartedly say: *"Yes! I am leading a life that I am proud of – in every way"*. And that feels so darn good. In large part I attribute my success to the many people in my life that I connect with regularly – it is such a great feeling.

Where do you start?

Where do you begin to get your life on track? How do you get a toe hold on something to anchor you? Here's what worked for me.

We are bombarded by stimulus in many forms: friends, family, co-workers, TV, web, emails and it can be overwhelming. The natural response is to try and control it, control others to make them see the right way to do things – our way. But just like traffic and weather, we really don't have control over ... any of it! Most people have figured out they can't change the weather but focusing exclusively on changing yourself is hard.

What You Can Change

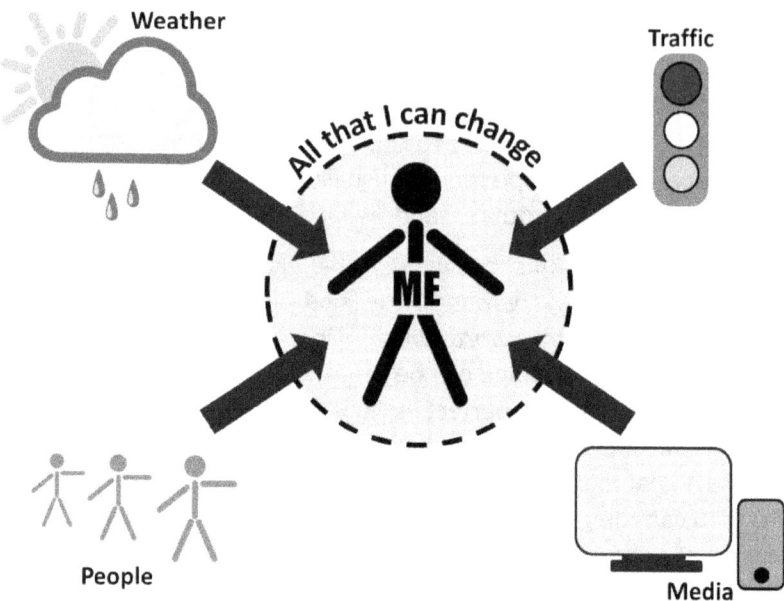

I realized that I have the power to change only two things: me and how I perceive the world. That means I cannot change other people, what they should do with their lives or what they think of me. (In fact, what they think of me is none of my business). This simple yet powerful idea is summarized succinctly in the Adult Child of Alcoholics Serenity Prayer[11]:

> "... grant me the power
> To accept the people I cannot change
> The courage to change the one I can
> And the wisdom to know that one is me."

Okay, so once I had that figured out, where do I start with me? Start with your head; for example, if your head is not fully committed to losing weight, no amount of dieting is going to work until your brain fully understands and commits to the goal of eating less. Our programming can be so clever to fool us — self-sabotage (more on this later). But we know when this happens because we feel crappy about ourselves.

After many months tackling this problem, it came to me that one may begin to achieve peace and serenity if one answers two questions honestly and thoroughly:

What do you want?
What do you fear?

For me, both questions were daunting. I did not feel worthy to answer the first and initially did not have the courage to address the second. It took work ... a lot of work, and help from others to explore these questions. Without dealing with these core issues, there is no lasting change or happiness. Figure out your answers because we deserve to get what we truly want and can say NO to fear. *"Stare fear right in the eye to steal its power"*[12]. Answer these questions without limits, reservations or critical analysis.

EXERCISE – You just won the lottery! (10 min)

Congratulations! You have won the Engineering Your Life Sweepstakes - $100M! And because of the magical nature of the experience, you suddenly have absolutely no commitments of any kind. That means no family, friends or co-workers trying to horn in on your good fortune. Furthermore, there is a complete media blackout on the event so nobody on the planet has any idea you are now a hundred-fold millionaire. Oh, and we have special dispensation from all governments so there are also no applicable taxes – it's all yours.

Now the question is ... what are you going to do with the money? Don't think too hard about it. Just set your timer for 10 minutes and start writing below how you're going to spend it.

If you are having difficulties doing the Millionaire exercise, why do you think that is? You may want to jump to the "Obstacles" chapter now and come back to this section later.

1. _____

2. _____

3. _____

4. _____

Initially I spent a great deal of time in "discovery mode" that is reading all about my issues without actually DOING anything – it didn't help. Stop getting ready – YOU have to step into action RIGHT NOW!

Go back and do the exercise.

Engineering a New Life: Where am I?

I would rather you give the book to someone who is ready to change rather than do a half-hearted read. Be bold and make a decision right now.

Go back and do the exercise.

What do you want?

Hooray and congratulations! I hope that you are now feeling a glimpse of joy. Read on to get more of them.

The previous exercise should help reveal what you really want without the limitations of time, money and responsibilities that usually block the view to your Nirvana. In my case, being a Motivational Speaker with freedom to do my own show is what I aspire to achieve. (If you're scheduled to attend one of my workshops based on this book, then it's become a reality – and thank you for your support!)

If your exercise reveals you want something like having parties on your own private island in the Caribbean entertaining travelers from around the world, showing them your amazing art collection, then maybe you should consider opening up a Bed and Breakfast. If on the other hand, your notes indicate that you want to hide out in a remote part of Mexico in a full security compound complete with distillery, you may want to jump to the Family Dysfunction section in the next chapter ... seriously.

Okay, you don't have $100M - yet. But now that you know what you really want, why not go after it anyways? Don't worry about obstacles (like cost, time, skill, education, permission or other practical considerations) we will address those in a later chapter. This chapter is about defining what you want. If you don't think, speak and act on what you want, things will just happen around you. Be proud of your life. Lead the life of your dreams! This book is only for you, so be truthful. Don't hide the reality of what you really and truly want from yourself and the universe.

So what do YOU want? What makes you feel happy? Consider goals that would be fun, cool and exhilarating. If you feel a smile forming on your face or in your soul, you're on the right track. State your desire in the positive; for example, rather than "eat less junk food" choose a goal of "eat healthier snacks". And don't worry about getting the perfect words, you can refine later. Right now just get your inspirational ideas on paper. Here are some examples:

- Better Job $80K+ that makes me jump out of bed in the morning
- Find my Soul Mate
- Be my own boss
- Be a home owner in 3 years
- Be in the best physical shape of my life
- Travel to a new place I've never been – every single year

- Sync expectations with partner on date night every Friday
- Smile to at least one person every morning on my way to work
- Make time for my kids every day – no matter what
- Design the world's first skateboard made of coconuts
- Track my monthly expenses and save at least $100 a month
- Be the best I can possibly be
- Peace, serenity and clarity
- Meet 10 new people a week
- Have a healthy sex life as soon as possible

EXERCISE – Life Goals (15 min)

Set the timer for 15 minutes and without judging or thinking too hard or worrying if it's okay, or if you can afford it, or if you have permission – write down at least 7 goals that you want. Use your pencil right now! Provide as many specifics as you can. The first two columns will be filled in a little later so ignore these for now.

RANK CATEGORY GOAL

A la Carte Goals

Need help picking some goals? Try the family meal approach used at Chinese restaurants. Pick one goal from each major life category: Career/Income, Romance, Health, Friends and Family, Spirituality, Home and Art depicted in the Life Wheel below. I include a category for "Art" because it can add so much joy, serenity and wonder to one's life. I highly recommend some artistic element to your list such as drawing, singing or just going to see a concert occasionally. A balanced approach with modest goals from each category can be an effective way to get started. And you can always change goals later if they are not serving you. Pick at least one goal from each category for a balanced life.

Seven-Segment Life Wheel

The "Home" category may seem unusual but we all have to live somewhere so I thought it would be good to dedicate some time and energy to make it a nicer place. Some ideas for goals include: finding a new one home, fixing up your existing one or maybe finally committing on paper to that kitchen remodel.

Where's fun? Fun can be everywhere! Have you ever noticed that good movies regardless of genre contain some humor? And that's not a coincidence because good movies mimic life. We need humor and fun every day! So there's no need to limit fun to any one life category.

I intentionally did not include a category called "Hobbies". The reason is that many hobbies are loner activities and that is not in keeping with the underlying theme of this book to connect with other people. So if for example you like scrapbooking, perhaps set a goal to meet with some friends or your mother to scrapbook together every Sunday night. If you like fly fishing which tends to be an individual sport, consider it a Spiritual goal to bond with the river once a month and make a point of not thinking of work or family issues. There is nothing wrong with time alone time; my advice is to do so in moderation.

You will also notice that there is no "Things I want" category. That is because things aren't really what you want. If you dig deeper you will find there is some other need you are trying to fill with stuff. For example, consider the desire for a new car. You may want a car to accommodate all the soccer and musical equipment needed for your kids' extracurricular activities; that's a Family goal. Or you may desire a new car because you seek to get promoted to the executive team and want to project an image of success and style – that's a Career goal. Do not fool yourself by saying: *"I want it because I want it!"* More than likely you are avoiding some issue that is holding you back. Just because you don't want to address the underlying issue does not mean it doesn't exist. If you find that honestly the reason you want the sports car is to feel more powerful, that's good to acknowledge; you may want to probe a little further to find out why that is really important to you and perhaps find other more satisfying ways to address those needs.

I was debating using 6, 7 or 8 categories for the Life Wheel and chose 7 in part because that number has been lucky for me (I was born on July 7th); also because there is an interesting correlation with Pythagorean number mysticism[13]. Briefly, this is the concept that numbers have inherent meaning attributed to pre-dynastic Egyptian culture.[14] Seven seems to be particularly appropriate for this book as it embodies the essence of growth. And if I may throw in, there are 7 notes to the western music scale. Once you have reached a point of health in the 7 categories of your life, you can make beautiful music!

EXERCISE – Life Goal Review (30 min)

If you did not write down seven goals in the last exercise, go back and do that now with your new insight. If you do have your goals set – excellent! Now check your goals across all 7 life categories. Start by entering the appropriate "Category" value for each goal in the previous exercise (Home, Romance, Career/Finances, Friends/Family, Spirituality, Health/Fitness and Art). You should strive for at least one goal per category. Consider changing or adding additional goals to cover more aspects to your life.

Reality Check

Okay, so now we should do a reality check on our goals. If your dreams will impact others, talk to them about it – partners, kids and friends. If they are not directly part of your goal, can you negotiate one of their wants to make it a win-win situation by sharing time and/or money requirements? Set yourself up for success by sharing your goals with the important people in your life; they may even help you in ways you did not consider. And they can be your very own cheering section!

Do you have the skills or knowledge to achieve your goal? If not, that's easy to fix - add in the necessary research, workshop or formal education needed to achieve your goal.

Is your goal timeframe realistic? Probably not, but that's okay too. Once you learn more about what is involved and how much time you can dedicate, adjust the goal dates. Getting a handle on your life is a process so your goals will change – and you've got an eraser in your toolkit.

If you're not willing to act on a particular desire, perhaps that is because it is not what you really want. For example, you may want a rocket-propelled dune buggy, but because they cost $150,000 you just shrug your shoulders and decide you could never afford it and therefore the only conclusion to reach is that life is so unfair. After some serious deliberation, you may find that what you really want is to have more fun and be included with friends. So for example, if you hosted a beach party with Karaoke, horseshoes, Mai Tai's, marshmallows by campfire and invited everyone you know, that may meet your true desire. And you could do so at a tiny fraction of the cost of the dune buggy.

Do you have some conflicting goals? Figure out which is more important to you right now. You can always shelf a goal for later consideration. In that case, I would recommend setting a date for review so you can re-evaluate where you are in your life and make any desired goal adjustments.

Explicitly figuring out what you want then making choices accordingly will free you from having a vague unsatisfied feeling or grumbling at the world. If you find that your interests have changed – no problem – change your goals!

Checking for Balance

The Life Wheel can also be useful to check in with yourself periodically and see where your life is out of balance and needs a little work. Green indicates "all good" – I have a goal in place and making good progress. Yellow means I have a goal but it's either not quite right or my progress is minimal and some focus is in order here. Red is well … a red flag. *Warning - attention needed here!* I may be neglecting this part of my life and from my experience that means denial because I just don't want to deal with the issue. Or it may mean that I've thought about it, have a pretty good goal but there is nothing going on – nada. In this case, a change in tactics may be in order; that is the activities that support the goal.

There are many factors within each category so it may be challenging to assign only one color. The idea is to get an average of that particular aspect of your life so you can get a top level snapshot of the whole picture. My colored my life wheel on 7/13/14 is shown below. (White, light gray and dark gray correspond to green, yellow and red respectively). For example, when I think of the state of my health, these thoughts come to mind: not needed to visit the doctor in the last year, no major pains reported lately, exercising regularly, feel good about my food choices – that's a green!

Author's Colored Life Wheel

As in life, if your seven life wheel areas are not in good shape you will likely not be happy and often thinking about the problem areas. It's like a bicycle some broken spokes – it doesn't ride smoothly. So the colored life wheel allows me to focus on the red area in my life. Heck, for the first time I'm seriously considering dating sites – yikes!

EXERCISE – Category Coloring (15 min)

Now it's your turn. Set the timer for 15 minutes and break out the crayons! Color the Life Wheel for yourself based on how you feel about each category. Use your medium of choice: crayons, markers or pencil crayons. Be sure to fill in the date so you can compare when you check back on goal status in 3 months. (If you mess up, there's some extra Life Wheels in the "Answers" section at the end of the book).

Status of my life categories on _____

Supporting Your Goals

Now that you have figured out some important, balanced goals for yourself – don't hide them in a drawer. Display them proudly in a prominent spot to remind you to work on them and celebrate progress. A weekly check is highly advisable especially if you have someone supportive with whom you have developed the goals.

Surround yourself with images and objects that remind you of your life goals and desired state of mind. For example, if serenity for 30 minutes a day is one of your goals, get a tiny Zen garden that can give you the hint to turn everything off or go for a walk. If travelling is what you seek, plaster the fridge with images of your dream destinations. Be your own cheerleader and take every opportunity to be successful!

We are Robots

Carbon vs. Silicon

I was discussing the potential of artificial intelligence with one of my friends; he insisted that could never happen because only flesh and blood are capable of intelligence and machines are just circuits and gears. I disagree. In my opinion, the materials used to create something are completely independent of their function. For example, computers were created long before the invention of the transistor in 1947[15]; the abacus first appeared as a Sumerian device over 4000 years ago[16]; the slide rule was first built in the 17th century[17] and used on the manned moon missions; and the Antikythera mechanism, an ancient computer that determined planetary positions including eclipses, was devised by Greek scientists in 205 BCE and found relatively recently in a shipwreck in 1902[18].

A robot by definition does not possess intelligence. The essence of a robot is that its behavior is completely predictable; it's not about what types of parts are used. Someone has designed or manipulated its every move. If you press a button, the robot will tell you the time; if you wave at it, it will say hello. And I believe that people can be robots. If the hand that waves is made of steel alloy and wires or bone and nerves, it doesn't really matter. What matters is determining who is driving.

Stop Driving on Autopilot

We Are Just a Story

Egyptians believed the heart was more important that the brain. The heart was weighed symbolically against a feather in the resurrection ritual to determine if you lived a good life. The brain on the other hand was unceremoniously mashed up with a stick and sucked out of the nasal cavities[19]. The heart as central to the self was common throughout many cultures including our own until medical science found out the heart was just a pump and the magic happens in our brains. (Symbolically, however I think the Egyptians had the right concept in terms of living with your whole heart – more on this later).

Brain research has come a long way and experts understand the mechanics that turn trillions of brain cells through neural networks into the illusion of "Me". All our thoughts, ideas, aspirations, beliefs, history, feelings, obsessions and sense of self are all there in approximately three pounds[20] of chemicals and electrical activity.

According to Neural Psychologist Paul Broks, our "self" is just the history of what has happened to our body - a story our brain tells itself[21]. This may seem at first like an unusual and perhaps unsettling idea. What you think or believe doesn't really tell the world who you are. You are defined by your actions - the part you play on the stage of life. "You" are the sum of every action you have ever taken.

However, "you" as a story is also powerfully empowering! Since we have experiences every day, the brain we wake up to in the morning is not the brain that we put on our pillow at night. Researchers confirm that contrary to previous belief, there is clear evidence that brain cells grow, die and modify during our entire lifetime[22]. So ... if our selves are a story ... WE CAN CHANGE THE STORY – ANYTIME WE CHOOSE TO DO SO! Don't like how you react to people or situations? Don't feel in control of your emotions? Want to be a better you? You can!

Why did I just do that?!!

My ex would ask me this question in frustration and bafflement with unnervingly regularly and usually adding "we just discussed this"; I could never provide a satisfactory answer. (Oh, how I hated those moments). She was right! I should have at least known the reasons for my actions.

When someone pushes your "buttons" do you tense up, shutdown or tell them why they are wrong or perhaps say something nasty in response? That's your programmed, unconscious responses. "Fight or Flight" was the original motivators behind these types of reactions which served us well in the past - you don't stop and negotiate with a hungry mountain lion. However, in the vast majority of interactions today, such extreme responses do not serve us – they undermine relationships. In my case, I had developed reactions to specific situations or "triggers" as a child that enabled me to survive my dysfunctional family. Unfortunately, these same reactions stayed with me as an adult even though they were no longer needed; they in fact became damaging to me and those I wanted close to me.

Do you ever choose to alter your daily routine? Can people predict where you will be, at what time maybe even what you will say? Do you find yourself playing on-line games, solitaire on your computer or watching TV for hours? Do you "checkout" regularly and find yourself at your destination without remembering the trip? Does your life seem like an endless stream of mundane required activities punctuated by desperate attempts to have fun? Still think you're not behaving like a robot at times? I sure did and really had no idea that my behaviors could be changed let alone how to do so.

Do you ever feel manipulated? It may be that someone has figured out your programming and is hacking into ... you! I remember the first time that I actively took charge of my reactions. The person who was pushing my buttons wore an expression that I had not seen before directed at me. I had set a boundary that clearly sent the message: "You have no power here". I was never again manipulated by that person – because of what I changed about me. (Details on how to set boundaries are covered later on).

Don't Let Other People Push Your Buttons

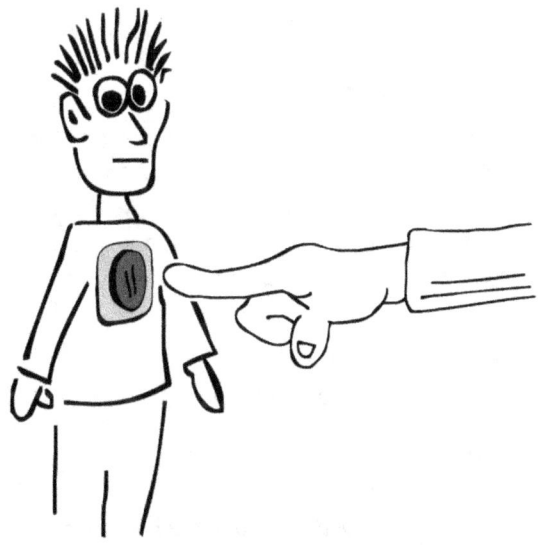

We are constantly reacting automatically to our environment. We have to – because the human body is complicated and there are many tasks vital to our survival and dedicated to our lower brain functions such as breathing, regulating blood pressure, scanning for threats, and filtering out irrelevant data[23]. Try the following exercise to see how you filter data without being aware of it.

EXERCISE – Count the F's (2 min)

Set the timer for 30 seconds. Count the number of letter F's in the paragraph below, then write down the total in the space provided. Check the answer in the back of the book when you are done.

> *The kings of Persia often exhibited FIERCE fighting skills if they were on a campaign to extinguish foreign foes or IF they were intent on engulfing their neighbors for the explicit purpose of EXPANDING their territory for prosperity and ADULATION of the fair masses.*

Total letter F's: _____

There are other ways in which we exhibit auto responses – just like programmed robots. We seek comfort foods when we are stressed. Some of us revert to childhood behaviors when our internal triggers are activated. We act with prejudice when stereotypes are unwarranted. We fabricate excuses when confronted with unpleasant facts. We will do almost anything to keep our illusions intact and not alter our comfortable routine even though it may be hindering or harming us.

Observe and Know Thyself

I can better recognize these automatic cycles when I make time to calm myself, notice the world and how I am reacting to it. Well, you can and you can do it whenever you want. It's the simple act of self-reflection, getting a little quiet time and focusing on what is going on with you. Many of you may at this very moment have readily come up with a few objections:

> *"Easy for you to say! I have to work!"*
> *"I've got kids – it's impossible for me to ever have that kind of luxury".*
> *"Yeah, someday but not now!"*
> *"Sure once I get through my list … if I have any energy left".*
> *"I would really like to … but I can't".*
> *"Maybe after the kids are in bed, I've done the laundry and … zzzz".*

If you had any of these responses at the ready or something similar, then you are indeed deeply steeped in a routine programming and need some help to step outside this cycle and examine what's going on with you. Are you uncomfortable or even afraid of what you will find out? If you truly want to engineer a better life for yourself – this is a required step. It may be difficult – but oh so worth it based on my experience. Be brave and do it for yourself!

If you have children, you can only help them by helping yourself. It's like the takeoff instructions given on every airline: *"In the event of a cabin depressurization, air masks will drop from the panels. Put yours on securely first then tighten the straps BEFORE assisting your children"*. Why? If you're not okay – you can't help anybody else.

How do you begin to observe yourself? First, you need to want to change. If you're not convinced of this, put down the book and try again in a few days; without that fundamental desire, you will not be successful.

You do want to change? Good. Next, figure out a good day and time that you can dedicate to a little quiet time and reflect on your day. Fill in the commitment below:

EXERCISE – Reflection (30 seconds)

For the next _____ weeks, I will reflect on my day every (circle):

Monday   Tuesday   Wednesday   Thursday   Friday   Saturday   Sunday

At the following time: _____   AM   PM

Signed: _____   Date: _____

No time like the present. Do your first reflection right now.

EXERCISE – Reflection Now (15-30 min)

Turn off the TV, phone and anything else that may be distracting to you. Soothing music can be beneficial if you have some. Tell your family members that you need this time undisturbed and why it's important to you; or just get away to your nearest coffee shop, park or beach. Pull out your journal (get a pretty/cool/fun one that you will want to open and use often) or a pad, your favorite pen and a beverage of your choice. Now just start writing:

- How was your day?
- How do you feel right now?
- What do you want right now?
- What do you fear right now?

All you came up with was a sentence? No problem – that's all that's in you today. Try again next time. Need more space? Great - get yourself a bigger pad. Do this for at least 7 sessions.

Life Outside the Matrix

Would you like unreasonable happiness, contentment and true moments of joy? If you still don't feel that you are entitled to being happy, do it for your kids - they are affected and you will help them by helping yourself.

Like the Matrix[24], you have a choice, either the blue pill where your life stays the same, you live isolated; or pick the red pill and you will embark on a journey of new life, unimagined joy, and unreasonable happiness connected with the world around you - the true reality of life. You have the strength, curiosity and spirit to take the challenge. Don't think about it - be in the moment. You need to do less thinking and more feeling. What do you want? What feels right?

Opportunities are everywhere!

I was on a brisk early January hike in Modjeska canyon (there was actually snow on the ground here in southern California) and chatting with Meagan. Turns out she is also a trainer and went on further to mention that she was struggling to fill a position in her group. I looked behind me and motioned to the guy following directly behind us who seemed distracted and looking intently at the trail. *"Brian! You may want to talk Meaghan about an opportunity"*. In a conversation just moments before I found out Brian had a big dilemma - either transfer to Atlanta with his current employer or find a new job as a trainer. So literally right in front of him was an opportunity to solve his problem. Opportunities can come from our actions/inactions, circumstances around us and other people; everything and everyone is interconnected.

Instant Gratification

I discovered Aldous Huxley's Brave New World[25] in a High School literature class. It is a dystopia novel where the writer depicts a London society in the year 2450 where shrinking the time from desire to fulfillment is paramount. He invented a drug called SOMA, the ultimate numbing agent. After ingesting it you felt wonderful, there were no side effects and you did not feel any compulsion whatsoever to make any choices. With this drug you were happy to live according to your programming. This novel written in 1931 was a warning. And his nightmare has come true with the web – much sooner than predicted.

With the help of the web, you can indulge in many mind-numbing activities: porn, games, gambling and shopping. Consider how easy and quickly you can fulfill your desire for ... anything. Go to your browser, type in what you want, what size, what color, enter your credit card information then click and it magically appears at your door often within 48 hours. When you get what you want in two days, the time to plan, yearn, hope and anticipate practically disappears robbing us of the joys of our imagination. For many people, the Web is in fact programming them like a lab rat! But you can change your programming. Use the web as a tool – don't let it use you!

Programming

Before we talk about human programming, let's look at a simple programming example that is familiar – a lawn sprinkler system. As shown on the next page, our watering system has a controller with three individually sprinkler lines for the front lawn, garden and backyard as well as a sensor to prevent wasting water while it is raining. We want to water at 8 am but only on Mondays, Wednesdays and Fridays and vary the length of watering for the three areas (15, 5 and 10 minutes respectively).

Water Sprinkler Programming Example

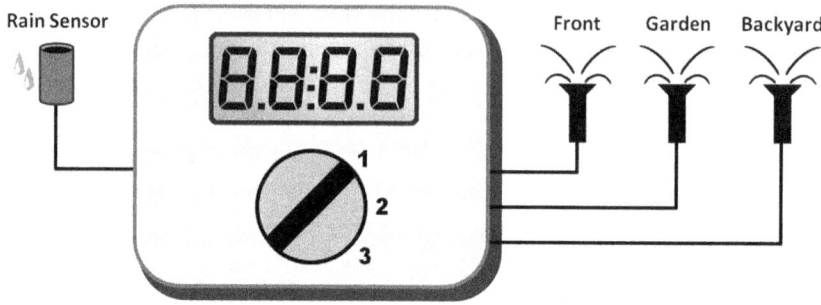

Inside the controller there are chips that make up a small computer with a specific task of determining when and how long to water. We have to do some programming so that our sprinkler works as we just described. On the next page is a flow diagram to illustrate the operation. Each box is a decision that the computer must make (yes or no) based on various inputs (day of the week, current time, active rain sensor and how long the sprinkler has been on). Each of the possible actions are linked to a box so that the computer knows what to do next. They're kind of dumb that way – you have to tell them everything! (Yes ladies, at times some of us males also need to be told every step in some situations, myself included).

Water Sprinkler Flow Diagram

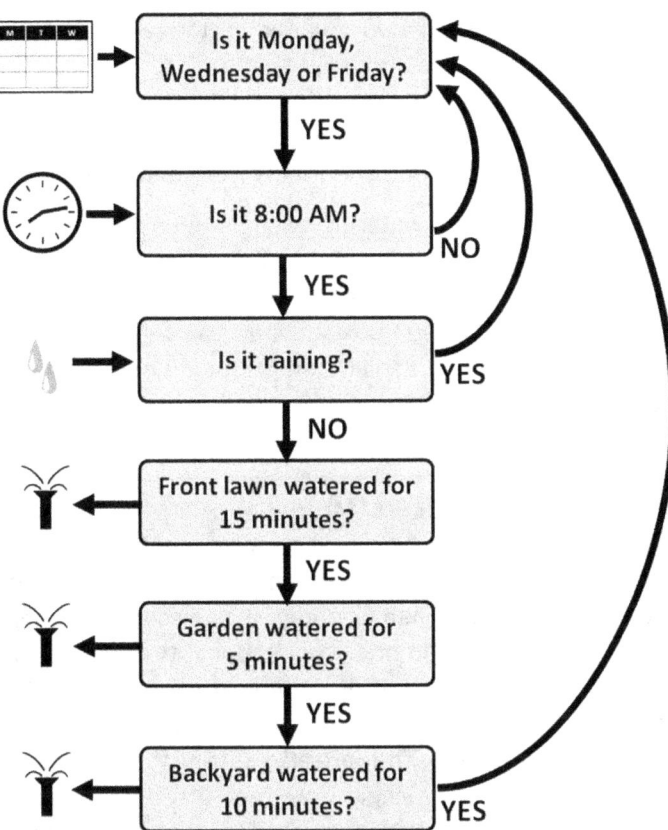

Starting at the first box, the computer uses the on-board calendar to determine if the day is a Monday, Wednesday or Friday. If the answer is YES it moves on to the next program element. But if the answer is NO then it loops on the date question patiently waiting for tomorrow; it does absolutely nothing else until this question is affirmatively answered. So let's pretend it is now 12:01AM on Monday – hooray the computer can move on to checking the time!

In the second box the computer must now determine if it is time to start watering. It's a simple process to compare the controller setting of 8AM with the current time using a special time reference chip. If it is exactly 0800 then programming moves on to the next box. If it is not, then it goes back to checking for the day of the week. We don't want it to loop back on itself because if you set the controller start time right after lunch on a Saturday, it would go off on Sunday at 8AM; and some of us like to sleep in on Sundays so that would be bad programming.

The third box now checks for rain. If it is not raining, then the computer moves on to start watering the front lawn. But if it is raining, then it will go back to checking the day. If it just waited for the rain to stop then the front lawn would get watered immediately after it stopped raining – not helpful. So the computer must methodically go back to checking the day, the time and once again for rain – then it would loop for day, time, rain – day, time, rain – until all three conditions were met in sequence.

In the fourth box we start watering and continue to do so until 15 minutes have elapsed. Then move on to the garden for 5 minutes and the backyard for 10 minutes. Once all three areas have been soaked, the computer goes back to the first box and starts all over again. It will get to the second box because it is still Monday. But it is now 8:30AM so the answer to the second box is NO and it will go back to checking the day. That's how programming works!

Human Programming

Do you know someone who says moments after meeting you: *"Oh, I will never remember you – I'm not good with names"*? This drives me crazy! We all like to hear our own names and greeting someone using it will instantly establish an authentic connection; the above opening comment does the complete opposite. This person is telling me from our initial interaction that I am not important because they are not even going to try to remember me. They are also putting themselves down by asserting this inability. Every single time they utter those words their programming weakens the synapses related to name retention. Instead they could open with: *"I am working on remembering names and you STEVE are a person I want to remember"*. I struggled with remembering names too and consciously began improving it. Listen to their name will full attention and if necessary, ask them to spell it. I try to use their name at least three times in that first conversation. The closed loop of hearing the name from your own mouth seems to improve retention. For special people, I will jot down their name with a follow up comment on my phone.

For a dramatic example of human programming, consider the intriguing loop that Mary Sue Campbell experienced in 2010[26]. One day while working in the garden she suddenly had the sensation that there was something "wrong" with the world around her. After being whisked to the hospital, she was diagnosed with Transient Global Amnesia – a condition in which you cannot form new memories. Her daughter Christine quickly arrived to find her conscious and alert in a hospital bed. The conversation went something like this:

Mary Sue Campbell Brain Loop

> INT. HOSPITAL ROOM - DAY
>
> MARY
> *What's the date?*
>
> CHRISTINE
> *It's Tuesday, August 24th.*
>
> MARY
> *My birthday has already passed?? Darn!*
>
> CHRISTINE
> *Don't worry we have it all recorded.*
>
> MARY
> *What happened?*
>
> CHRISTINE
> *You were working in the garden …*
>
> MARY
> *… that is so creepy!*
>
> CHRISTINE
> *I called the **paramedics** and they …*

When Christine utters the word "paramedics", Mary's pupils became hyper dilated and she returns to the top of the script: *"What's the date?"* Mary repeats her questions again completely unaware that she asked them 90 seconds ago. She repeats the same dialog not once or twice – she repeats the exact same dialogue 75 times to an exceedingly patient daughter – for two and a half hours! The amazing part is that she repeated the same words, in the same order, in the same way including the emotion behind the *"Darn!"* at missing her birthday.

The doctor who had seen a few similar cases believes that Mary's verbatim behavior loop was due to her brain responding to a limited number of stimuli (hospital room, doctor, daughter's voice, equipment beeps) in exactly the same way every time her brain reset – just like a machine. The identical inputs provided to the brain produced the predictable outputs in the body. Mary's questions were not unusual and in fact everybody would react the same way, with their brain going into "survival mode". I believe that in our daily human interactions we need to get out of survival mode in order to break out of our own loops.

After a few hours Mary's loop began to expand in duration with old memories slowly creeping back until she returned completely back to normal.

In some respects, we are not so different than the sprinkler system, we are just more complicated. We have five senses to take in a vast array of inputs and we have many more options on how we respond to situations. We filter the data stream because there is simply too much to take in. Normally our brain blocks out the irrelevant stimulus but on occasion the throttle to our memory retention is thrown wide open. Such a situation occurs when athletes are "in the zone" or if you perceive that you are in mortal danger. David Eagleman had such a childhood experience when he fell off a roof[27]. As he was making the 12-foot drop he imaged that his experience was like Alice falling down the rabbit hole – time slowed to a crawl. (The actual time was a mere 0.86 seconds.) He suffered no serious injuries but the adventure turned into a question that he would investigate many years later. He believes that the apparent long time to fall is a trick of our memories. Because our body knows this is a life threatening situation, the brain stores all incoming information looking for something to survive the experience. The texture of the tiles below, shape of the clouds, smell of the nearby Jasmine bush, guy with the blue shirt across the street … everything is stored and remembered in vivid detail. Perhaps the commonly reported feeling of "my life flashing before my eyes" is part of this state where the brain is madly searching our past memories for something useful to save ourselves.

Knowing how our filters work and consciously tweaking them to suit our goals is a highly useful skill. Consider a typical work day flow chart depicted below with the questions we ask ourselves based on inputs around us, actions we take and the sequence we go through – every single day.

Typical Daily Routine Flow Chart

[Flow chart: What day is it? → Sat/Sun: Go Back to Sleep; Mon-Fri → Am I late for work? → NO: Go Back to Sleep; YES → Is it raining? → YES: Take Umbrella; NO → Does the car have gas? → NO: Go to Gas Station; YES → Is there traffic? → YES: Freak Out!; NO → (end)]

Do you see the similarities with the sprinkler programming example? Acknowledging that at times we do behave like robots is essential to take control and begin to consciously re-program ourselves.

Making Choices

The opposite of acting like a robot is to exercise choice. Choices give us power and fulfillment. If you are unhappy in your life, I suspect it is because you are not consciously making meaningful choices. When we are consumed by the trivia of everyday life, it is easy to get overwhelmed with making decisions: paper or plastic, cash or credit, chicken or beef, skirt or pants, vanilla or chocolate.

You have a choice. Actually, you have a lot of choices. In the above example the robotic response is to get up if it's a weekday or roll over and go back to sleep if it's the weekend. To break out of the programming you might choose to get up an extra hour early on Tuesday and have a nice breakfast for a change or go for a walk to get relaxed before your presentation at the office. Every day you could actively decide what you wanted to do and not rely on your programming.

What you eat every day, every meal, every bite is entirely up to you. How you spend your time and with whom is within your power. You make decisions how and where you spend your money. You also can choose what you think about the world around you and what you tell yourself about ... you. Aware or not, good or bad, you make choices every day. I have found it is a whole lot better if I make conscious choices not default robotic ones.

"But I don't want to make all the choices!" Okay, then your life is whatever happens. However, be aware that you are nevertheless making unconscious choices. For example, when you sit in your cube at the office always staring at you screen, avoiding eye contact and not engaging with co-workers, you are telling everyone to stay away. And quite possibly you are sending the message to your boss that you are not interested in advancement opportunities.

> *"If you think you can do a thing or think you can't do a thing you're right".*
>
> Henry Ford

The power of choice is that you're never wrong. No matter what else happens in the world or what other people say, your choice is always valid because that is what you decided. Your decisions do have consequences and you need to consider what is likely to result from your decisions. However, you will never again be wrong regarding your choices. It's simply not up for discussion.

For example, let's say you decide to buy a set of noise-cancelling headphones at ACME Electronics. A week later you may see them advertised at Widgets-R-Us for a lower price. There's no need to regret the extra money you paid because you chose what you wanted and when you wanted it. You did not elect to do extra research or wait another week to find a better deal. There is value in having the headphones sooner and freeing your mind from thinking about the buying process. Time in this case to do other things was a higher priority than saving a few dollars. You can smile with your decision and bask in the glorious serenity provided by your new headphones. And of course you could now choose to return them and purchase at the lower price. (Mine had a 30-day full refund policy).

The difference in choosing is that you are making a wide awake decision and not relying on your programming that says for example: *"I must always spend the least amount of money possible."* This default thinking could in our example have turned out quite differently. Say you didn't buy the headphones at ACME Electronics and waited a week, a month, spending every moment of your free time to scour the internet, scanning the weekend ads for a better price. You find a bargain basement site that has a similar pair at a much lower price. And you get them. But they don't quite work (turns out they're not really noise-cancelling), you get frustrated, return them and are bitterly disappointed by the whole process: *"I never really wanted them anyways"*. You wasted a great deal of time most of which was spent in anxiety and unhappiness. And for the cherry on top of your disastrous experience, you're resentful when your friend tells you how much they are enjoying their headphones from ACME - which you recommended.

Take responsibility for your life – it's your turn to decide!

EXERCISE – Power of Affirmation (30 min)

If you have no problem making choices, skip this exercise. However, if like me, you at times struggle with feeling okay with making choices for yourself - read on.

Affirmations are a powerful tool to empower you to feel better about yourself and help you move to action. Review the following affirmations and repeat each one out loud three times. Circle any of them that particularly resonate with you. Or better yet, with your pen write your own affirmation in the space provided. (Feel free to mix and match pieces from the suggestions. Keep it short so it is memorable like a slogan or tag line).

- I CAN DECIDE
- MY OPINION MATTERS
- I HAVE THE RIGHT TO MAKE CHOICES
- TODAY - I AM IN CHARGE OF MY LIFE
- I DECIDE WHAT I WANT

- I CHOOSE FOR ME
- I TAKE RESPONSIBILITY FOR MY LIFE
- RIGHT NOW I CHOOSE _____
- _____
- _____
- _____
- _____
- _____

On a sheet of paper write at least one of these affirmations in big letters approximately as tall as your thumb is wide. Write with a pen or marker because you completely believe these affirmations to be self-evident for you. Make three copies for each affirmation. Cut out each affirmation.

Find three visible spots in your home, car or office to place these that you can easily see many times a day. Some places I've used include: nightstand, taped to bathroom mirror in view when brushing my teeth, refrigerator, above handle on door that you use to leave the house, car speedometer and computer monitor.

Now, whenever you see your affirmation, say it energetically out loud three times! You may feel silly; it may seem like it's not working – keep doing it! Like osmosis, someday soon you will notice that you honestly and actually believe it.

Software Debugging

When debugging software (i.e. finding the problem), a common diagnostic technique is to plaster the code with "print statements" - collect data at different points of the program to get clues as to what is going wrong. You can do the same thing with your life. It is essential to collect data dispassionately and objectively over the course of a few days or weeks then look for the pattern. Don't try to analyze as you collect the data because you will influence the results which will not help you figure out what's going on. (More about this later in the "Scientific Method" section).

Nature of Human Interaction

Nothing is truly free; everything is an exchange of one sort or another. But this is not a bad thing. It's just the way life works. I believe all human interactions are such an exchange: money, time or desired feelings (not always positive). So be aware and choose for yourself if you are willing to make the exchange. Here are some examples:

- At a car dealer, you pay money and get a car
- At a restaurant, you pay money and receive food and service
- At a wedding, you bring a gift and get included in a special life moment
- At the hospital, you bring flowers to comfort a friend and get an IOU for a reciprocal visit
- A total stranger holds open a door for you, they feel uplifted by your grateful smile

There's always a reason you spend time with particular people. If you are unhappy in a relationship, find out why. Discovering your motivation is crucial to change your programming and your life. It took me a really long time to figure this out for myself. To dramatically illustrate this point, consider the 1962 Cuban Missile Crisis[28] where the United States and Russia were on the brink of nuclear war.

Offensive missiles sites were under construction in Cuba within close proximity of US cities. The super powers were in a deadly serious relationship crisis that could have led to a global nuclear holocaust. Robert Axelrod, Professor of Political Science at the University of Michigan at the time was tasked with simulating various programs to discover the best strategy to deal with the crisis[29]. He invited programmers across the US to submit entries and every program competed against every other one 200 times. Although many detailed approaches were proffered, some with thousands of lines of code considering many factors, the winning strategy called "Tit for Tat" had only 3 lines of code:

> **Line 1: Be nice**
>
> **Line 2: Echo your opponent's move**
>
> **Line 3: Occasional random kindness**

What this means is that in the opening interaction, you are friendly to your opponent but if he is aggressive you retaliate in kind. You do not however, attack mercilessly and assume they are your enemy forever (i.e. don't hold grudges). Rather, you give them an opportunity to act differently and respond in kind – good or bad. This seems like a good approach to me for personal relationships as well.

When you seem stuck in a loop, doomed to keep repeating the same behaviors over and over and over … stop! And make a better decision. If you're not sure what to do, just do just one thing differently; no matter what it is, break the routine and see what happens. If that new thing does not help, try changing something else until you get closer to the result you want. Like the cold war program above, a small algorithm change can break you out of your loop. Don't be a robot!

Obstacles

In this chapter we will look at what may be holding you back. Your goal here is to figure out which obstacles apply to you. Each topic heading contains three boxes to help you start building your roadmap.

Check the box for each heading appropriately. Use a pencil since you may change your mind later. In the next chapter each of these obstacles will be discussed with suggestions on how to walk around, jump over or eliminate them.

Don't Have Time ☐ yes ☐ no ☐ maybe

What is important to you? Don't fool yourself. If you say six-pack abs is important but you have to eat that piece of chocolate cake because it smelled so good at the bakery, a sculpted body is really not what you seek.

We spend time on what is important to us. (Or at least what our programming has deemed so). Finding out where you actually spend your time will help you determine what is truly important to you. You can then consciously decide to change priorities and reallocate time accordingly. If you feel time is an obstacle for you, do the next exercise.

EXERCISE – Time Analysis (30 min)

Complete the table below for all the major activities in your weekday. You can round off to the nearest half hour. Having difficulty determining where you spend your time? Fill in the exercise mid-day and at the end of your day. If your days vary considerably, or if you want to factor in the weekend, perform the analysis for each day of the week using the charts provided in the "Answers to the Exercises" section at the end of the book.

Time Analysis Chart

Time	Activities
6AM	
7AM	
8AM	
9AM	
10AM	
11AM	
Noon	
1PM	
2PM	
3PM	
4PM	
5PM	
6PM	
7PM	
8PM	
9PM	
10PM	
11PM	
Midnight	

Daily Summary	
Categories	Time/Week
Sleeping	
Working	
Commuting	
Eating	
Hygeniene	
Partner	
Kids	
Friends	
TV/Web/Email	
Total	**24 hrs**

Life on Autopilot ☐ yes ☐ no ☐ perhaps

Do you ever feel like you are going through the life motions without any emotions? This state of mind or lack thereof is often called numbing. It's complacency. I used to do this all the time; grab a pint of ice cream and sit in front of the TV or computer for hours and not feel anything or engage with the world. I lost so much time of my life to numbing. It feels good to get out of the gloom of a monotonous, unengaged life.

Limited by Commitments ☐ yes ☐ no ☐ maybe

We all have commitments but do you feel they limit your future choices? If you can discuss, re-prioritize or re-negotiate existing commitments, this is not likely an issue for you. If you feel trapped by some agreements (spoken or implied) maybe this topic is worth exploring.

Lack of Confidence ☐ yes ☐ no ☐ maybe???

Tough one but really Important. Like posture, you can practice balancing a book on your head to train yourself to stand erect, or you can feel better about yourself and you will naturally stand taller without thinking about it. Diets, makeup, new clothes are fixes from the outside. I have found that working from the inside is much more effective and longer lasting. Once your core is aligned on the path of your dreams, the rest flows out inevitably and joyfully!

Self-Sabotage ☐ yes ☐ no ☐ hmmm

Banish negative thoughts because they don't help you. If you tell yourself that you can't – you won't. It all starts in your head. Since you can imagine anything you want, why not paint a glorious picture of success? This is in fact a common technique used by Olympic athletes before big competitions[30]; winners know their biggest competitor is in their head. When you see a track star in their logo laden warm up outfit with their eyes closed, they are most likely visualizing the race in their head; their feet slide effortlessly into the starting blocks – hands poised on the line – they are the bullet shooting out of the gun – flying down the track propelled by the crowd's jubilations – soaring through the finish line – raising both arms to the sky in gratitude for their win. After this exercise, they open their eyes, take off their sweats … and just do it.

One of the most annoying phrases I hear when the topic turns to any form of art is: *"Oh, I have no talent for that."* How do you know? If that was said about non-art activities some people would never survive. At some point you had no idea how to drive a car, shop for groceries, tie your shoelaces or even walk. I don't think too many kids get away with *"Oh I have no talent for taking out the trash."* Did someone help you to learn to drive and effectively shop at the local market? You can get help and learn to paint with water colors or play the harmonica. We humans are amazing when it comes to adapting and learning new skills. It's all about the stories you tell yourself.

If this applies to you, consider adding a creative Life Goal that you have always wanted to do. I firmly believe everyone should have some art in their lives – it's so right … brain (more on this later). Then once you have picked some art you fancy, start thinking about how you can learn more; who do you know with some experience? Or just get some supplies and start playing. At the writing of this paragraph I decided to go out to buy masking tape, spray paint and put a blue wave on my car! Resistance to creating art is often a lack of confidence. Don't worry that the art you do may not be good enough; art is to express who you are and communicate it to others. Yup, it's more of that human connection stuff.

Let your inspiration take you on a ride!

Addictions ☐ yes ☐ no ☐ just maybe ...

One description of the mechanism of addiction is a short-circuited brain that goes to a default mode to preserve itself and survive because it has detected something has gone terribly wrong. Like a defective computer, your brain goes to its basic programming and displays an error message. Problem is ... our error messages can be hard to diagnose and correct.

If you have an addiction, or think you might be controlled by something – get help. In trauma cases you check for breathing, stop the bleeding then deal with the hangnails. With engineering a new life, dealing with addiction is the critical first step because until that is under control, nothing else matters. There are over 50 different 12 step support programs that can help, requiring only a token financial contribution. You will find people with the same addiction that understand what you're going through; some are well along the road of recovery to light the way for you.

Co-Dependent Relationship ☐ yes ☐ no ☐ ??

Are you in a healthy relationship? The following questions may help you determine if you are:

- Can you say NO to an activity or request?
- Are you respected?
- Do you have control of your time and finances?
- Can you spend a significant amount of time with other people?
- Do you enjoy being with the other person?
- Do you look forward to seeing them when you're apart?

If you can answer yes to all of these without hesitation, most likely you have a healthy relationship. If not, you may want to explore the areas in question. Pay attention to those issues, check your feelings, observe your reactions and see how long they last after an encounter. A diary was useful for me to record my observations and examine trends. Engineers love data. And data serves a useful purpose – you must measure it to improve it. Think of a marathon runner who wants to shave 30 seconds off her time. She tracks her time, diet, hours of sleep, state of mind every day to learn what factors are contributing to reaching her goal and those that are not.

Quest for Perfection ☐ yes ☐ no ☐ maybe so

For decades I believed that my never ending quest for perfection was the highest calling on earth. Part of my programming was the belief that you were a better person if you achieved perfection. I remember practicing the accordion one day when I was 8 years old proudly belting out a polka with gusto. After I finished, my father shook his head dismissively and said: *"There were two mistakes"* and walked away. I tried again but with considerably less enthusiasm.

As an adult I would often be consumed with trimming the lawn perfectly, organizing the tools in my garage perfectly, planning the most efficient route to do errands perfectly. At work I would aspire to create the perfect spreadsheet (with color), the perfect training program and the perfect website. And any comment that wasn't completely glowing would be interpreted by my programming as failure. Employers love perfectionists because they are so driven to get things done with the highest quality and finest attention to detail. The biggest challenge supervisors have with perfectionists is convincing them to take a break or a vacation to prevent burn-out. Seeking perfection seems like a good trait but it really is not. That is because you get so consumed by the task that you sacrifice other aspects of your life: family, friends, health and fun. I am not advocating mediocrity; setting a goal of "greatness" or "reaching your best" is inspiring and it leaves room for being human. Most of the time the standard for "perfection" is set by someone else and it is far better to choose your own standards.

There is also the darker side of perfectionism to consider – constantly seeking approval. I used to seek perfection in order to get the kudos and pats on the back. Sadly, doing things perfectly (or not doing them at all) was the only way I would feel good about myself. It seemed that I was only as good as my last great achievement. My self-worth was completely defined by what I did.

Dysfunctional Family ☐ yes ☐ no ☐ not sure

How do you know if you have a dysfunctional family of origin (the family you grew up with)? Are you still being affected by your childhood experiences? A few trusted people suggested that I had some issues with my father's alcoholism many years ago and sadly I dismissed them all. Denial ran deep with me. If you have even the slightest hint that you may be affected by such a history, *please, please, please read the following paragraphs*. For me, this discovery was the key to my recovery and my new life. Do you spend a lot of time ruminating thoughts in your head? The rest of the world often sees this as a vacant stare. I know that place very well. Complete the exercise below to help determine if dysfunction affects you.

EXERCISE – Determining Dysfunction (10 min)

Check the items below that apply to you:

☐ Do you isolate and prefer not to engage with others?
☐ Do you avoid experiencing bad feelings?
☐ Do you seek to control every aspect of your life?
☐ Do you frequently feel you are unworthy or not good enough?
☐ Do you feel guilt when you stand up for yourself?
☐ Do you fear intimacy or being abandoned?
☐ Are you plagued with self-doubt and struggle to make decisions?
☐ Do you often believe that your opinion doesn't matter?
☐ Do you find it difficult to accept criticism or compliments?
☐ Are you frightened by conflict, angry people or authority figures?
☐ Do you often feel responsible for situations and other people?
☐ Do you judge yourself harshly and think that's a good thing?

If you checked more than three boxes, you most likely have some family dysfunction issues. But do not despair because you are not alone. Like many others, I suffered from most of these and now rid of all of them – and you can too.

It was astounding for me to find out that many of my behaviors as a 50-year-old adult stem directly from the traumas I experienced when I was five. To hacksaw through those chains, I had to go back and understand early childhood experiences from an adult's perspective. Groups like Adult Children of Alcoholics or Dysfunctional Families (ACA)[11] can help.

All the above issues are manifestations of abandonment. With a little hindsight, I can now see that my family of origin dysfunction was the key to understanding my insecure and self-defeating behaviors that led to a life that I did not control. I was truly programmed by the forces experienced in my childhood – and didn't even know it. With this new found knowledge and getting outside my own head have enabled me to discover what I want in my life then making choices to turn them into reality – this book is that journey.

Believe me, I understand if you are thinking right now: *"Oh no, not me ... I can handle this ... there's no way I could talk to anyone about this ... I'll be fine ... I just need to figure it out on my own"*. Here's my last and most heartfelt suggestion for you. Go to one ACA meeting. All you need to do is walk through the door, say hello, take a seat and just listen to the childhood stories that may be a lot like your own. It sure seemed that way to me at my first meeting. If it doesn't work out, you wasted an hour of your time. It may however shed some light on your behaviors and lead the way to peace, serenity and happiness. It is of course, up to you. Refer to the "References" section in the back of the book to locate an ACA chapter[31] in your city.

Anything Else?

The obstacles mentioned are the ones that I have experienced directly or indirectly through others. Do you have other issues? If so, jot them down in the exercise below. Like a storm of ideas while you're trying to sleep or attempting to remember 5 grocery items to buy (or was it 6?), I have found it useful to write them down; when those thoughts are on paper or on my phone and out of my head, I am more relaxed and able to enjoy the moment before me. Get all those obstacles out of your head now. Then you can smile and know that you have contained all your issues on paper – a critical milestone in engineering your new life.

EXERCISE – Other Obstacles (20 min)

You may have other obstacles that I did not encounter. Give it some thought and jot down any possibilities you discover.

1. _____
2. _____
3. _____
4. _____
5. _____

Second Opinion

I found it difficult at this stage to be certain what my issues were. If you feel the same way, you may want to consult with a trusted friend, family member or therapist to get their opinion by reviewing your checkboxes with them.

Fear is a scary thing.

What do you fear?

I was at a library recently and the themed book display caught my attention. The poster read: "What are you scared of?" with the expected graphics of spiders and snakes. Other titles included: The Exorcist, America Bewitched, Rats, Area 51 and ... Trigonometry for Dummies! I am seriously tempted to write a chapter about Trig because for me it's such a beloved topic that I would like to share that with you, perhaps in the sequel.

Some Fear Manifestations

Fear - we dread it, try to hide it, deny it and even dress it up – Halloween. Here are some common fears:

- Public Speaking
- Flying
- Heights
- Bugs, Spiders, Snakes, Bats
- Embarrassment
- The Dark
- Intimacy
- Death
- Failure
- Rejection
- Birds
- Commitment
- Pain
- Uncertainty/Unknown
- Needles
- Dirt & Germs
- Not being good enough
- Strangers
- Dogs

Fear of public speaking is high on the list I believe because it is related to a number of subtle fears such as embarrassment, uncertainty, rejection and failure. I have given a great many presentations to audiences of all sizes and have personally experienced each one of these fears and have lived through them all! I now relish any public speaking opportunities and embrace the unknown aspect as a great way to experience the diversity of people and situations. Occasionally I get brief moments of fear: *"Yikes! I have to talk about self-sealing stem bolts next week and I have no idea what they are!"* Such a fear is a powerful motivator to start researching and preparing for my talk. The important thing is to face the fear and deal with it.

In some cases, fear serves a vital purpose indeed – it helps us survive. The mechanism to detect dangerous situations and react quickly has been an essential trait for our survival as a species. For example, imagine you are hiking in a lovely forest, enjoying the beautiful day – when suddenly you spot a snake! For most of us, the first reaction is to freeze, evaluate the situation (is it poisonous?) then continue on or run the other way. (Personally, I skip the evaluation step and just get the heck out of there). This fear is a healthy one that originated a long time ago when snakes where a serious threat to our existence. Today, because hospitals with anti-venom serums are within quick reach of most population centers, death by snakes is rare. In the United States 7,000 to 8,000 people are bitten by a venomous snake each year with typically less than 6 that are fatal[32]; you are 8 times more likely to be killed by lightning rather than die from a snake bite. Our fear instinct may well contribute to this low statistic by minimizing encounters with these creatures.

Other primitive fears no longer serve us, and can even hinder us such as fear of open spaces; in the distant past, humans were vulnerable to attacks by fast moving predators, especially in open spaces. But today, those animals are endangered and have minimal interaction with human populations. However, some individuals still have this fear and it severely handicaps them as their reclusiveness makes it difficult to have normal social interactions. These primitive fears which are now obsolete are also known as phobias and have clinical names like: Acrophobia (fear of heights), Astraphobia (fear of thunder & lightning) and Agoraphobia (fear of markets … just kidding, it's actually the fear of situations where escape is difficult).

Because these fears are hard-wired from way back in our evolutionary history, it is difficult for our brain to distinguish between real and imagined threats. Both of these can generate the same dose of adrenalin to get us to react instantaneously to preserve life and limb. Unfortunately, some of these false fears can cause serious side effects that impact our quality of daily life.

Engineering a New Life: Obstacles

There was a little girl who was hounded by a recurring bad dream about monsters. Her father told her the next time she had the dream to turn around and look at them – and she did! She bravely looked right at her demons. She noticed that one had 5 purple eyes, another with a funny nose and ... suddenly they weren't quite so scary.

Like that little girl, look your monsters in the eye. Write down your fears in the exercise below and examine them naked on the page where they lose their power.

For example, you may have fear of dying on your list. Well, fear not because you – like all the rest of us will die. Perhaps what you're really concerned about is leaving your children unprepared or that you will not have done everything on your bucket list by the time of your departure. Ok, those things you can do something about. Start a trust fund, update your will, set a goal to enjoy time with your daughter every weekend and teach her a life skill like cooking or rock climbing. Plan an international vacation with your partner every 6 months and display posters of your destinations.

While the fears are in your head they are nebulous, paralyzing and can even grow like a cancer choking the joy out of your life. Get them out of there! Put them on paper now.

EXERCISE – Monster Exorcism (30 min)

Set your timer for half an hour. This is a challenging exercise that you may need to revisit a number of times. That time however will be well spent. Write down at least four of your fears below. What is the worst possible scenario that you can imagine happening for each of these fears? What do you think is the actual issue behind your fear? I've provided an example to give you an idea of what to write.

FEAR: *Public speaking*
Worst Case: *People will laugh and think less of me.*
Actual Issue: *I need to work on my self-confidence.*

FEAR: _____
Worst Case: _____
Actual Issue: _____

FEAR: _____
Worst Case: _____
Actual Issue: _____

FEAR: _____
Worst Case: _____
Actual Issue: _____

FEAR: _____
Worst Case: _____
Actual Issue: _____

FEAR: _____
Worst Case: _____
Actual Issue: _____

FEAR: _____
Worst Case: _____
Actual Issue: _____

FEAR: _____
Worst Case: _____
Actual Issue: _____

Uncertainty

Fear of the unknown or uncertainty is shared by many of us. I have moved from fearing what will happen to embracing it. Recognizing that change is the natural state can be liberating. It is the concept of permanence that is the illusion. In nature, everything changes; it is only a matter of time scale that makes it deceptive: plants, animals, seasons, landscapes, planets, suns. Why then would other aspects of our lives not be equally subject to change (like our jobs, relationships or our lives)? Embracing the uncertain means you no longer worry about what will happen; no longer try to control it; no longer fear what will happen. Rather, look at what is happening right now with wonder and joy!

Changing Behaviors

90 Second Rule

Kelly Clarkson sings a powerful message in one of her songs: "Get your hands off my trigger"[33]. And what I have discovered is that there is no need to tell people to do so as I have the power to disable the trigger firing pin myself. No one can control my feelings or behaviors unless I let them.

This idea appears in many places, the most impactful for me is the 90 second rule coined by Jill Bolte Taylor, a neural anatomist[34]. From brain research conducted over the last 10 years, scientists like Jill have a much better understanding of how our heads work.

We are constantly receiving a barrage of inputs from the world around us. Our brain then interprets this data, matches it up with our experiences and beliefs which then triggers physiological responses like increased heart rate, flushed face or that "pit in the stomach" feeling. For example, if someone cuts me off while I'm driving, my brain will receive the visual information of a car suddenly careening right in front of me; it will then process it with my memory of another such event a week ago, add in the fact that I am late for work and then send a surge of adrenalin through my body that will ... propel me ... To Get Irritated ... AT THIS *@#$%^*!! WHO IS INTENTIONALLY RUINING MY ENTIRE DAY!!!

Sound familiar? Here's the amazing thing that Jill discovered; the time for the entire process just described (getting cut off, body responses, internalizing the event, throwing out a few obscenities) and for the adrenalin to completely flush out of my body, returning to my previous calm state ... *is only 90 seconds*.

So why is it that when I get to work, I'm still railing on about the social miscreant that was completely responsible for putting me in this foul mood to anyone that will listen? It is because there's a feedback loop in my head that recalls the event and starts the same cycle in motion getting me more indignant about the incident – over and over and over. In the past, I would perpetuate this loop with comments like: *"Well I can't help feeling this way – HE CUT ME OFF!"* The reality of the situation was likely that he too was running late, completely oblivious to me and did not have a dastardly plan to sabotage my day. My big lesson was that I can stop this loop; I have such an opportunity every 90 seconds.

The 90 Second Rule to Change your Behavior

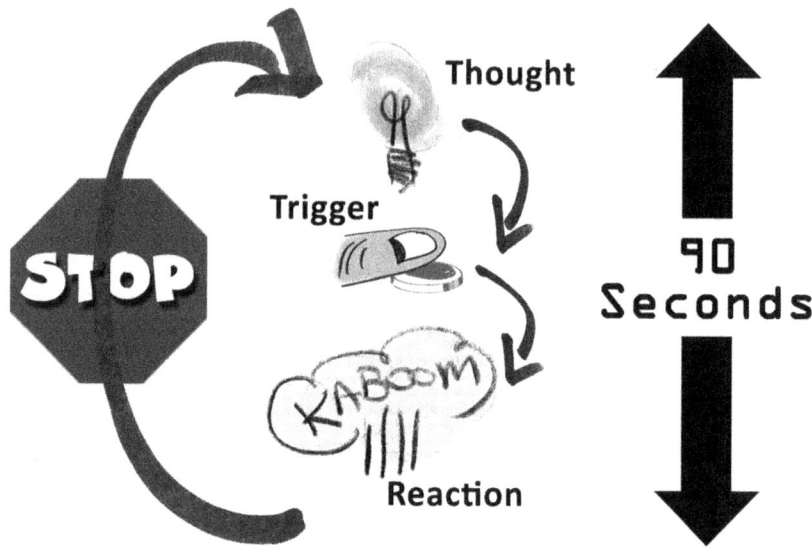

Now armed with the 90 second rule, I can retake control of my day and my life. I can choose to have a different thought when the feedback loop begins to form. I may not catch it in the first 90 seconds – but that's okay because there's another opportunity in a minute and a half. I no longer blame events or other people for how I feel. I have found that a good way to break the 90 second cycle is to talk positively out loud. The path from mouth to ear to brain seems to be more effective than just thinking the thought[35]. Talking seems to engage a different part of our brain enabling better analysis and retention.

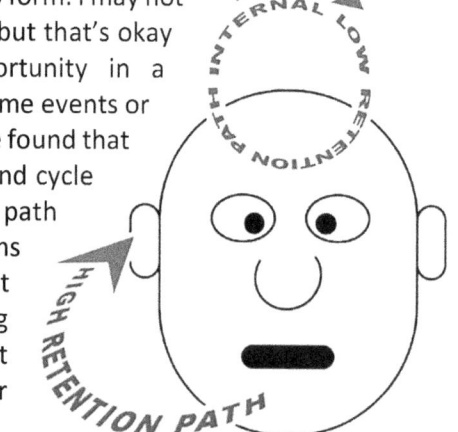

Do you recall Mary's memory loop example discussed earlier? That was also 90 seconds before her brain reset and repeated the exact same series of questions for her daughter – interesting.

Accountability
Another option to saying words out loud to yourself is to share it with someone. Someone you just met or an acquaintance can be a treasure trove of unbiased feedback. And it is an opportunity to connect with new people. When you verbalize a thought or your intent for the day to someone, it gives the idea substance; and with a regular person, it enables accountability because they will likely ask you for an update the next time you see them. For more sensitive issues select a trusted friend or family member. My therapist and my ACA fellow traveler served this role well to keep me moving towards my goals.

Every 90 seconds you can choose a different course of thought, an alternate direction; you can break the loop and direct your thoughts to what you want and actions that support your life goals.

Conscious Living
Being fully engaged in your own life is a powerful concept. To me it means being aware of what is around you at this particular moment then decide for yourself what you will do next. If you are bored with your life, ask yourself this question: *"I wonder what will happen next?"* Then look around for something that interests you ... and pursue it! For example, before you crack open another beer or one more soda, consider the impact it will have on your health goals. It's not about guilt or to restrict what you can do; it's about a choice. In this case I can see two of them. One, do you really even want this soda? (You certainly don't need it). If this is your programming in action – break the loop! Second, which do you want more? Beer right now to get that satisfied feeling in your belly or the beginnings of a 6 pack at the end of the month? (Okay, for some of you the wait may be longer).

Who You Need to Change
You don't need to save the world or feel guilty because others have it worse than you do. Just make your corner of the world and the people in it better by the end of the day. Helping others helps you. However, your job is to change the only one you can – you. Even if you have children, your job is to take care of them, advise them and guide them – they are ultimately responsible for making their own decisions.

Judging Others
We are quick to judge and categorize. For example, when I meet someone "What do you do?" comes up pretty quickly in the conversation. We want to peg a person to understand and place them in our world view, not to mention rank them relative to ourselves and

others we know – the pecking order. This is practical, yet impersonal and boring. I used to do it; I would ask people's profession to see if I could relate to them. When asked, I would identify myself as an engineer. I would take great pride in declaring it for the success it implied. But now I get annoyed if people respond with: *"Ah ... an engineer"*. Because now I know that I am way more than that!

Such an approach steals a richer inquiry and discovery of the people you encounter. Be curious about others and yourself. Yes, it will take longer – but you will enjoy the process; the "To Do" list can wait another day. Enjoy the moment now and you may even make a new friend.

Small Increments

A tiny change over a long period of time can have as astounding impact – either positive or negative. An extra dessert a week can mean having to upsize your pant size in only a few months. Or, foregoing one cracker or piece of bread at every dinner can mean you need to invest in *smaller* clothes in a few months. The key is to make your change a habit that is of negligible discomfort or inconvenience in the short term but make it a permanent change. Here are some other examples to consider:

- Pick up one piece of trash a day that you didn't leave
- Say one kind thing to a stranger every day
- Wait 5 seconds before responding in person (phone or email)
- Take the stairs once a day instead of the elevator
- Thank your significant other for being in your life
- Ask your child how they feel right now
- Do one thing extra per week that will help your boss
- Admire one sunset a week with a loved one while hugging them

Dealing with Numbing & Making Time

I really did not want to waste any more time in my life so I decided to get rid of my TV and home internet connection. It's amazing how much time I now have for other fun and constructive activities! If you plan such drastic measures, pick replacement activities that you really enjoy. For me surfing and playing piano did the trick.

Dealing with Commitments

Making and keeping commitments are a part of societal membership; it is the fundamental requirement to create and maintain relationships. When you say you will do something, people expect that you will come through. If you don't meet your commitment, most likely they will exclude you from their circle and you will lose a connection — the complete opposite of what we all seek.

For the longest time I didn't understand this. Many of my dysfunctional behaviors (seeking acceptance, projecting what I think others wanted and not standing up for myself) were rooted in this fear of losing connection. Now that I understand this need, I can take steps to establish and keep healthy relationships, like making reasonable commitments that line up with my life goals. If circumstances change, which they invariably do, I proactively communicate the situation and renegotiate the commitment. For example, something as simple as "David, I'm sorry but there's an urgent situation that has come up and I can't make dinner tonight, can we reschedule to Saturday instead?" was beyond me. I felt as though any commitment had to be honored at all costs; such a black and white attitude served no one, and I would feel bad or resentful for days — it was such a waste.

If I am honest, completely open and sincere, respectfully declining to make a commitment or renegotiate a previous one is actually pretty easy from my experience. I did have to figure out that it is okay for me to say no. That is wrapped up in confidence.

Getting Confidence

Self-esteem, feeling worthy, pride — all of these are synonyms for confidence. For a really long time I lacked confidence. There would be occasional moments of joy that would shine through. I remember being on a roll one evening unexplainably feeling really good as a late teenager and was making my whole family laugh — it was like being a stand-up comic feeling the audience and delivering exactly what they needed. Sadly, these moments were oh so rare. My demons would quickly regroup and chide me into submission: *"You are not worthy – shut up!"*

So how did that change? Well, as lame as it sounds, it was quite simply planting a marker in the ground figuratively and just proclaiming: *"I am worthy!"* And believing it — unconditionally. (Daily affirmations were instrumental in making this happen. More on this later). No reason or justification is necessary. We all are worthy just because we are here. After peeking out of my own shell, I made an astounding discovery: *I am not unique.* I am not special in a good way — that is to say anything I can say, do, make, create, sing, imagine … can be done and probably has been done by many others before me.

BUT ... the flip side is also true – I am not special in a bad way either! I have viewed myself to be different from everyone else on the planet – an outsider, someone not worthy to be a part of the human race – that is also untrue! There are many that feel that way about themselves. Don't believe me? Ask a few people, you may be surprised at the membership to this club. It is part of the human condition to think we are apart and separate when we are more alike than we ever dare to believe. And that makes us one big family. So chill out because those horrible negative thoughts you have about yourself, well, there's probably been at least a few million fellow humans that have had those exact same thoughts. It's okay, it's part of the process – and now it's time to move on, grow up and just be yourself with everyone else. It's cool – really.

Oh, and the other discovery that boosted my confidence was silencing my loudest critic ... me. Sound familiar? So stop it already!

EXERCISE - Confidence (15 min)

List all the things you do and put a star next to the ones you do well.

_____

_____

_____

_____

_____

_____

_____

_____

You have many reasons to feel confident, yet at times you may not believe it. So surround yourself with supportive people. You can choose to associate with people you like and respect and avoid or minimize time with people you don't. It took me a long time to figure this out. I wanted everyone to like me no matter how they treated me.

Lack of confidence made it difficult for me to deal with angry people. The root of my issue was wrapped up in developing survival skills in my dysfunctional family as a child. You have probably heard of the animal fight-or-flight response[36] described by Harvard Medical School physiologist Walter Bradford Cannon in 1915. It is a physiological reaction when something potentially dangerous to an animal or human is detected. For example, if you are walking in the woods and come across a large animal, most people will either attack it with whatever is at hand to use as a weapon or run away. These same auto-responses appear in most subtler forms during interactions with other people. In a room where yelling erupts, some people will immediately seek out the source and challenge the offender; others will turn tail and quickly leave the room.

There are actually two more options when faced with a threat: freeze and fawn. Freeze is the proverbial "playing dead" – do not move, do not speak until the threat goes away. I learnt this behavior as a child to deal with an alcoholic parent. By freezing, I would less likely be noticed and they would stop their tirade or attack someone else. However, this reaction was not effective for me as an adult. I used to do this when an argument got out of hand. The other person assumed that I was intentionally ignoring them and it only made them angrier and prolonged the argument. This is a good example of why it is important to figure out how you are programmed and change behaviors that are no longer useful or even detrimental.

"Fawn", the fourth survival response is total capitulation to the aggressor. The person under threat submits by agreeing with the instigator and allowing abuse in hopes of ending the conflict. I have unwittingly used this strategy in the past; for example, I would agree to do all kinds of tasks just as long as they stopped being angry with me. And then later I would become resentful of these commitments under given under duress. This is common behavior in co-dependent relationships and devastating to self-confidence.

Have you ever tried responding to an aggressive or angry person in an unexpected way? I've tried this – the reaction is amazing. They expect you to butt heads with them or cower; when you do something else, it throws them off their game and they are not sure what to do. That is your opening to have a civil discussion. You have just interrupted THEIR programming. I was only able to start doing this once my confidence was sufficiently secure. And confidence builds with each success you have. "How we react and feel about others is a mirror of what we think of ourselves".[37]

Dealing with Self-Sabotage
You are in charge of what you can do or say. So, if there's something you don't like about yourself, you can fix it by following one simple

suggestion: get out of your own way. Not trivial to implement but acknowledging that the biggest power and the biggest hurdle is YOU starts the process of looking at your life a little differently.

In my opinion, a subtle form of self-sabotage is the utterance of the phrase: "Everything happens for a reason". I am not fond of this expression because it abdicates responsibility. Instead of making choices in their lives, people who frequently use this phrase are shrugging their shoulders and are really saying: "there's nothing I can do. I'm at the mercy of my circumstances". Yes, everything does indeed happen for a reason – we live in a causal universe which means a result follows a cause. For example, if a soccer ball is kicked in the house, something will get broken. When you arrive home and see a lamp shattered on the rug with a soccer ball nearby, I sincerely doubt you would say: *"Everything happens for a reason"*. More than likely, you would figure out which of your beloved children was working on their indoor goal corner shots and do something to change the outcome in the future. Why not use the same approach for your life goals?

In the fate view of the world you are an observer, a reactor, going along for the ride, the one and only path – whatever that may be. And if you think that you are not on this perfect path, it's so easy to feel like a failure and miserable and wondering why things are not working out for you.

Life in my experience is much more fluid, dynamic and fun. If you're open to the possibilities of the universe and choose the one that is right for you right now, then you become engaged, active and an actor in your own life. If you live in the moment, these possibilities will make themselves visible to you … all the time.

So many doors are available to you – pick one!

Scenario 1 – Absent Adam

Bob wakes up to a day of possibilities. He grabs a banana from the kitchen fruit bowl on his way out the door. Standing tall, he greets people as he strides purposefully to the subway station. He savors his banana while appreciating the sounds, sights and smells on his way. He pauses, looking for a garbage can. Then walks over to a storefront and puts the banana peel in the waste bin. He resumes his walk crossing Main street disappearing into the crowd.

Moments later, Adam appears walking in the same direction. Adam is a little sad. He's been looking for someone with whom to share his life but has not yet found her. Truth be told, he is seriously thinking of giving up - that no such person exists. He's distracted by his thoughts. Not paying attention to where he's going, vaguely aware that he's going to be late for work. Yet he's complacent - doesn't want to really change what he's doing. After all, his life is okay; many people have it much worse. Everything happens for a reason.

As he crosses Main street, a woman crosses his path and stops in her tracks. Eve sighs wistfully as Adam continues on his way, completely oblivious. He ambles down the street without the slightest glimmer of the missed opportunity.

Scenario 2 – Awkward Adam

Bob wakes up. He turns his head to check the clock on the nightstand and is greeted with a shooting pain in his neck. But he is dedicated so rolls himself out of bed and gets ready for work. He takes a banana off the counter and heads out the door 7:45 sharp, as he does every day. He is understandably distracted this morning and refrains from moving his head sideways to manage the pain. After finishing his banana, he can't see the garbage can and in his aggravated state uncharacteristically tosses the peel over his shoulder just before reaching Main street.

Adam appears moments later. As he walks to work, he's distracted by his thoughts ... and doesn't notice the banana peel. He slips. Loses his balance, careens into the intersection and crashes into someone. But not just anyone. The most beautiful woman he's ever seen. He is not worthy. He is so embarrassed. He wants to just curl up and die. Eve gets up on her own. Waits. Adam blurts out a string of mumbled apologies. Eve shrugs and goes on her way.

About an hour later, Adam figures out how he could have handled the situation better. He will think about this missed opportunity and berate himself for a long time.

Scenario 3 – Active Adam

Bob wakes up. He turns his head to check the clock on the nightstand and is greeted with a shooting pain in his neck. He rolls out of bed, gets ready and is out the door with daily banana in hand. He is managing his pain the best he can and drops the banana peel after eating it - just before reaching Main street.

Adam appears moments later. He's been looking for someone to share his life with but has not yet found her – maybe today's the day. He's distracted by his thoughts ... and doesn't notice the banana peel. He slips. Loses his balance, careens into the intersection and crashes into someone. It's Eve. He jumps to his feet and offers his hand to help her up. *"My humblest apologies. Are you okay?"* Eve reaches out her hand and annoyance melts into delight as she sees who bumped into her. After a brief conversation. They both continue on their way. Adam has a wide grin ... and a luncheon date.

EXERCISE – Drawing (15 min)

Set your timer for 15 minutes. Grab your favorite pen or marker (no pencils) and duplicate the pattern you see in the left rectangle below to the one in the right. Do not use erasers in this exercise; move confidently forward with no concern for stray lines. If one square is not quite right, that's okay; forget about it and move on to the next square.

Use the grid to help you clone the image by focusing on one square at a time shading the black regions as defined in the original. After doing a few squares I think you will find that it's fun!

Now turn your picture upside down. See? You can do art! Didn't think you could be an artist? Of course you can. And by this example, I hope the doors to the many possibilities open to you are beginning to unlock.

Dealing with Addictions

No way around it, you have to deal with your addictions. I have some experience with 12 step programs and they are amazingly helpful. A term used in these programs is "reaching bottom". I really understood what this meant when I was so ashamed of my life that dying seemed imminent and almost welcome. But with help I came to realize that compared to that awful place I was in, anything else – embarrassments, disappointments or failures were trivial. If I live another day, a get another shot at getting it right.

Over-eating can be an addiction – right there in plain view. It was the case for me. When I was avoiding feelings or some tasks, irritated, bored or lonely – I would go eat something. It would drive me crazy knowing that there was ice cream in the freezer or chocolate in the pantry. I would resist for maybe a few days but then I would have to devour it all.

I took care of the cause of the problem and food no longer held any power over me. Then it was simply understanding what I was putting into my body and how much. I can now eat what I want and leave the rest. I sometimes think of what is left as an offering to the gods. My mother would cringe at the food waste. Better yet, I have learned to take or make smaller portions.

A good first step in dealing with an addiction is to avoid people and places that are related or remind you of the unwanted behaviors. This circumvents some of your programmed brain responses or triggers.

Dealing with Controlling Relationship

If you are in a co-dependent relationship, some corrective measures may include: clearly articulating your needs, changing your expectations and setting aside time to honestly discuss what each of you wants in the relationship. I would recommend doing so in a private, pleasant and quiet place such as a park or beach away from interruptions and distractions.

More extreme options to consider are professional relationship counseling or time away from each other. It's amazing the clarity one can achieve (speaking for myself) when you take sex out of the equation and have time away from your significant other. From my experience, the worst thing in my relationship was trudging on in old established patterns, avoiding dealing with the real issues and abdicating my responsibility.

Negative Thoughts

Negative thoughts are a different thing. In 2014, the average adult in the US spends 590 minutes consuming media and is exposed to 362 ads[38]. Curiously, this is not as big an increase as I would have expected; those same statistics are 309 and 340 respectively for the year 1945. However, that is a great deal of information every single day primarily from advertisers looking to sell us something. The strategy often uses the formula:

1. You need more X in your life.
2. If you buy Y, life will be perfect and all your dreams will come true.
3. You must get it RIGHT NOW for only $Z.

To get your attention, a big part of the ad will try to convince you there is something lacking in you or your life. It is rarely true or necessary but their job is to create a need for their product or service. Unfortunately, that feeds into many of our vanities, weaknesses and insecurities.

It's not the billboard, magazine or web ad that's the real problem. It's the messages that get trapped in our heads. Because we have the ability to replay these over and over and over ... we can cause damage to ourselves and obscure the exit to sanity.

I had an amazing experience that illustrates this idea. It started as a small thing, a slight rejection. As result, I was not feeling good about myself. I avoided my default programming – which used to be reaching for a mega-caloric comfort treat. Instead, I talked it out – faced it head on. I felt better but still sad and disappointed. A little while later I decided on a whim to send an email to an old friend that I had not contacted in a year to give him an update and thank him for his role in my recovery. Afterwards ... amazingly ... I felt back on top of the world! Why? Because I got outside my own head and thought about someone else. The next time you're having a pity party, or just don't feel good about yourself, try changing your focus to someone else.

Dealing with Perfection

I learned to let go of perfection and enjoy what can happen. For example, consider planning the perfect date or dinner. If it goes perfectly, you may get a short burst of satisfaction. But you likely will have focused and spent a great deal of energy of each and every detail to attain perfection, constantly comparing your expectation with what actually happened. And forgetting about what is important, connecting with your guest.

If you let go and are open to whatever happens, it can be fun and exciting. If it goes wrong, there will at least be a great tale to share later. And what a great opportunity to show your date/partner how you deal with adversity – a far more attractive trait rather than executing a perfect evening.

I used to torment myself with making perfect decisions: *"What is the best use of my time today? Should I take the new job? Should I buy this house?"* Then after making the decision, inevitably, I would find at least one reason to beat myself up for making the wrong decision.

Embrace imperfection! It's the cracks in life that make it interesting. When you stop chasing perfection, you are free to soak up the moment and be ready for fun, opportunity and joy. The more I get into these moments, the easier it is to stay there. And when I do fall back into old ways, I don't like it and motivated to get back on the high road.

Dysfunctional Family of Origin

Many of the obstacles discussed (numbing, addiction, confidence deficit, negative thoughts, quest for perfection) have at their root a troubled childhood. For me this turned out to be THE central issue that held the key to my recovery.

One evening when I was at my grandmother's house while my parents were on a date night it struck me that drawing a portrait of her would be fun. I was probably 8 years old. I remember becoming fully immersed in the process – looking at her intently while my hand traced lines quickly and confidently. I was so proud of the result; my grandmother smiled at my enthusiasm and proclaimed: *"You're an artist!"* (This was big for her too since she rarely smiled). When my parents returned I showed my masterpiece to my mother. Her eyes bulged out of their sockets and she was furious! She told me that my drawing was inappropriate and promptly tore it up. It seems I had a talent for clearly drawing every wrinkle and it made her mother look ancient. I was hurt, confused and ashamed for making my mother angry. As a child, my response to this and other similar events was to engage less with the world and instead withdraw into my own world that I could control – much safer. It would be 25 years before I would attempt another sketch.

Looking back as an adult with clarity, the event makes perfect sense. During that time growing up there were serious conflicts between my parents and my grandmother. Most likely there was an argument in the car on my parents' return trip to pick me up; my artwork provided an outlet for her frustrations. By creating an unflattering portrait, I was disrespecting my grandmother and making my mother look bad. With emotions already primed, my actions lit the powder keg and a nasty argument between the three adults ensued.

I remember a lot of yelling, crying and I was forgotten. They could not civilly handle their relationships so a child was certainly not a priority. I can now see how these poignant interactions with my parents were instrumental in writing my programming of self-doubt and abandonment.

If you have a similar childhood story, you may want to consider attending an ACA meeting. At first I dismissed it because it is a 12 step program. I used my religious view as an excuse not to explore ACA for a few decades and truly regret not doing so sooner. The program only requires belief in some power greater than yourself. The program is welcoming of all no matter what your higher power, whatever that means to you. As an example, some people consider the group presence in the ACA meeting itself a form of higher power – the collective ideas, perspective, support and understanding of the people in the meeting room are indeed more powerful than you alone. If you can believe in that, you are ready to go! (And of course you have the freedom to change your higher power at any time).

I highly recommend checking out ACA. If you need some support, bring a spouse or caring friend; if they are close to what you're experiencing more than likely you will both benefit. Hearing stories that you can relate to is so comforting and there are tools that really work to completely redo your life with happiness. In fact, the core of this book is about my journey discovering these truths - from an engineer's perspective.

The ACA Laundry List[39] is a great litmus test to see if you will likely benefit from the program. See if you can relate to the issues below that hinder adult children. Here is a simplified version of the Laundry List:

ACA Laundry List Summary
- Feel isolated and afraid of people
- Constantly seek approval from others
- Scared of criticism or angry people
- Drawn to compulsive personalities
- Victim perspective, attracted to other victims
- More concerned about others than yourself
- Feel guilty when defending yourself
- Addicted to excitement
- Confuse love and pity
- Difficult to express your feelings
- Judge yourself ruthlessly
- Have low self-esteem
- Terrified of abandonment

When new people attend their first ACA meeting, it is amazing to see their shocked reaction: *"How can this book possibly know my intimate story? I thought I was so utterly alone in this deep dark hole that nobody else could possibly understand."* That is exactly how I felt.

People recover at different rates, respond to the program differently and need different things. Find a group that feels comfortable to you. I suggest you check out at least six meetings before committing as the first few encounters can be overwhelming.

Family dysfunction is a powerful force and so insidious as it impacts your life every single day. If you are ready, please check out ACA - peace, serenity and joy await you.

A New Direction

Challenging Assumptions

If you are near a desk or table, knock on it – bonk, bonk! Does it feel solid? That object is actually made up mostly of empty space. And the stuff that is there is in constant, pulsing motion. Allow me to explain ...

In the early 1800's John Dalton[40] also rapped his knuckles on a table and wondered what it was made of at its most fundamental level – its structure. He proposed that the wood of the table and everything else consisted of tiny, tiny, tiny spheres - atoms. He believed that these atoms were the smallest bits of nature that could not be created, divided or destroyed. Instead they just recombine in a myriad of different ways in fixed ratios. For example, take a test tube of hydrogen, one of oxygen add a spark and poof! They will generate a few drops of water – H_2O; that's 2 atoms of hydrogen and one atom of oxygen. (By the way the process works backwards too; pass an electric current through two probes immersed in water and little bubbles of hydrogen will appear at one terminal and oxygen at the other).

Let there be water!

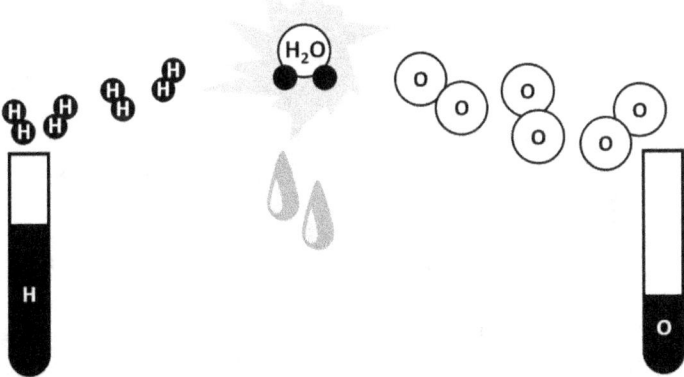

Other scientists expanded this idea with many experiments to determine all the different types of atoms: Hydrogen, Helium, Lithium, Beryllium, Boron, Carbon, Nitrogen, Oxygen, Fluorine, Neon ... and eventually discovered all the elements that collectively are called the periodic table. Every "thing" on earth can be seen as different combinations of these atom flavors. The periodic table was a great success by predicting elements as yet undiscovered because of blank spots in the table.

Periodic Table of the Elements

H																	He
Li	Be											B	C	N	O	F	Ne
Na	Mg											Al	Si	P	S	Cl	Ar
K	Ca	Sc	Ti	V	Cr	Mn	Fe	Co	Ni	Cu	Zn	Ga	Ge	As	Se	Br	Kr
Rb	Sr	Y	...														

A hundred years later in 1911, Earnest Rutherford[41] performed a historic experiment that opened the door to our understanding of what was inside the atom. He challenged the central assumption that the atom was solid and the most basic unit of matter. Rutherford placed a thin strip of gold foil in the center of a cylindrical detector screen. He chose gold because it can be stretched thin and it does not react with other elements. A magnified top view of the gold foil is shown on the right.

Peering Inside the Atom

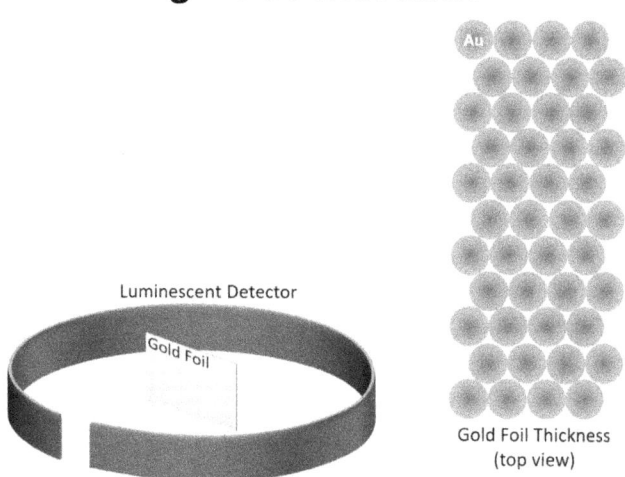

Gold Foil Thickness
(top view)

He then fired a high energy beam (alpha particles) at the foil target to see what happens. Specifically, where the particle would appear on the detector after hitting the gold foil would provide clues as to the internal structure of the atom.

Peering Inside the Atom - Setup

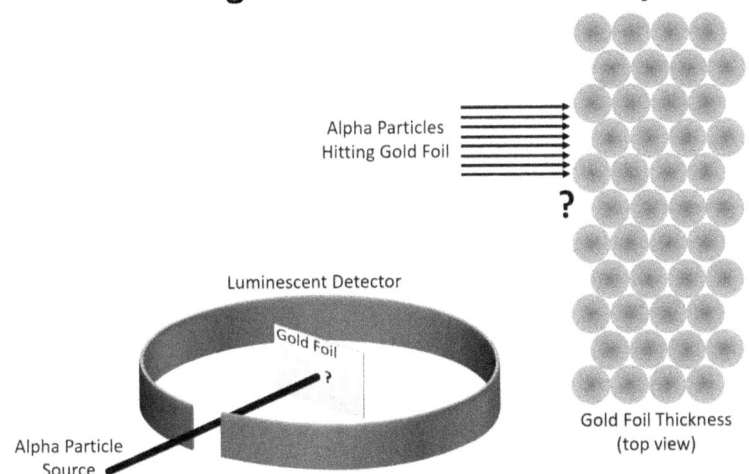

He expected that most of the particles would bounce straight back to the source with some deflecting at acute angles relative to the incoming beam. His plan was to measure these angles to figure out the internal structure of the gold atoms.

Peering Inside the Atom - Experiment

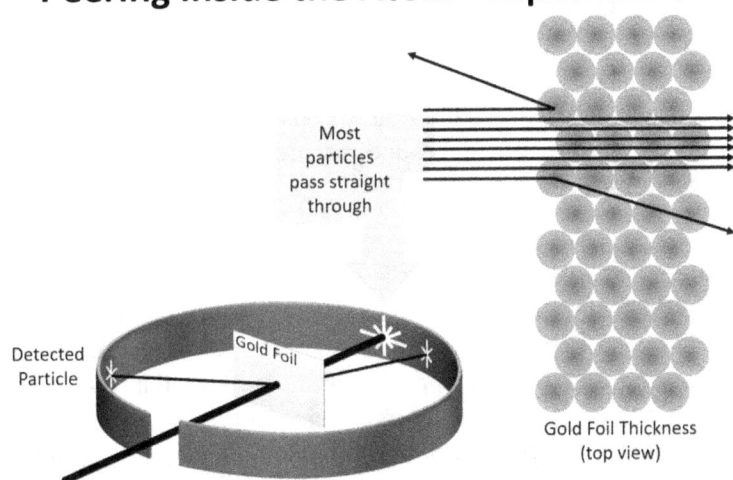

What actually happened astounded him and the world: 99.9875% of the particles passed right through the gold foil! Only one in 8000 particles was deflected. He discovered that gold atoms (and every other type of atom) are made up of a miniscule toffee center with the rest of the atom empty space. The reason the atom has volume is that there are a few gnat-like particles inside zipping around defining an atomic cloud. These particles were discovered a few years earlier – electrons. (There are 79 electrons in a gold atom. A hydrogen atom has only one).

Peering Inside the Atom - Conclusion

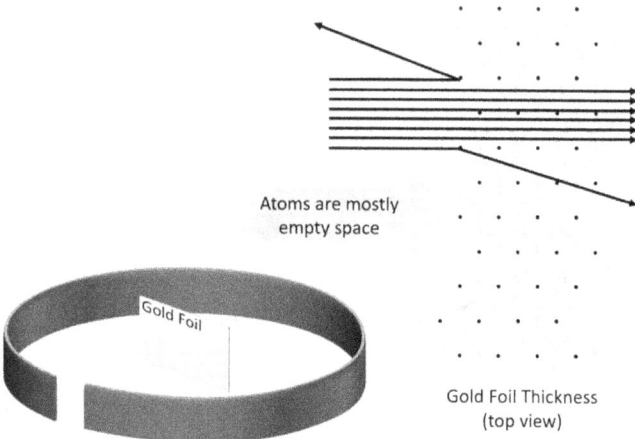

The diameter of the nucleus is 100,000 smaller than the diameter of the atom[42]. To give you an idea of the relative size of these three constituents of the atom, imagine an atom the size of planet earth - the cloud defined by the electrons. Proportionally, the nucleus would be only 1,000 feet in diameter. Think of the Eiffel tower at the center of a giant ball the size of the earth – that's the size of the toffee center! And the electron on that scale would be the size of a basketball! So atoms, which make up everything around us are balls of highly energetic particles and mostly empty space (which by the way includes you and me).

Like Rutherford, I've challenged the assumptions in my life and discovered that it had big holes in it. After completing some experiments, I discovered that one of the insidious gremlins that undermined my life was … assuming. I assumed myself into a corner. In order to be happy, I had to be perfect. I needed to say and do the right thing. And if I did not, then I would be bad and reprimand myself for the failure often remembering other such failures in endless self-recrimination.

Think of an invisible path somewhere before you that represents perfection – the one and only correct way forward. Many times I would be paralyzed to take the next step for fear that it would not be on that perfect path – and then I would be a failure.

First, I assumed that the daily goal was to be perfect. It also seemed to me that everybody else was on this righteous path and I was only one who was struggling so therefore should follow their lead. Most of the time I felt bad about myself and utterly baffled as to how to proceed in life. And this applied to my relationships as well.

Instead of discussing a difficult situation with my ex or asking a question, I would figure out the entire scenario in my head then present her with the outcome. After all that work, of course I assumed that the effort would be appreciated - it wasn't. Needless to say that added to my confusion and frustration.

Making assumptions limits your choices and keeps you inside your head – a dangerous place. If you're stuck in your head, you could easily be repeating an endless loop of negative thoughts: *"Why did I do that? That was so stupid. I can't do anything right. I better not say anything."* It's a bad place to be.

Try to get a bigger perspective and identify the assumptions you are making in your life. I find speaking out loud helps to shut off the tape. Your ears seem to be connected to a filter that eliminates absurd thoughts like: *"I am not worthy"*. Or better yet, discuss your thoughts with a trusted friend. They will have a different perspective and can usually give you a clearer picture of your situation. I would advise you seek a solution rather than getting mired in a complaint conference when you share your thoughts.

Once you have the assumptions fettered out, examine them and see if they are true. Are there better alternatives? Which line up with your life goals? Pick one and move in that direction.

I used to drink two glasses of wine at dinner. Didn't give it much thought, just became a routine part of the meal. Did I enjoy it? Selecting wines at the store was kind of fun. And I looked forward to that first sip to see what flavors I could detect. However, I must admit that there were times that the numbing effect of the alcohol was useful to get through my day. Although headaches the following morning were rare, drinking really did not support any of my life goals. Drinking did not take a great deal of time except when I consider the fuzziness that it promoted. I assumed that drinking was at worst a neutral factor in my life. However, I realized that when my thoughts were clouded, it was easier to get distracted and not focus on the activities that were important and joyful to me.

After I found my goddess, who rarely consumes alcohol, I made the conscious choice to remove alcohol from my menu. It was more important for me to be better connected to my partner and improve my health. Examine the routines in your life to challenge your own personal assumptions.

EXERCISE – Checking Assumptions (30 min)

Now it's your turn. Set the timer for 30 minutes. List 3 activities you do regularly and answer the questions below for each one.

Routine #1: _____

Do you enjoy it? _____

Why do you do it? _____

Which goals does it support? _____

How much time per week? _____

Better way? _____

_____

Routine #2: _____

Do you enjoy it? _____

Why do you do it? _____

Which goals does it support? _____

How much time per week? _____

Better way? _____

_____

Routine #3: _____

Do you enjoy it? _____

Why do you do it? _____

Which goals does it support? _____

How much time per week? _____

Better way? _____

_____

Scarcity and Abundance

I grew up in scarcity[43]. My grandparents were immigrants and they instilled the peasant mindset in their children who in turn passed it on to me. When I was a young teenager, my mother would reprimand me if I didn't spot and immediately pick up a penny on the ground. Her justification was: *"What are you going to do when you go to the store and you're short a penny?"* I argued that it did not seem like a good way to think about life. Years later I still have to say that she was wrong. It is far better to look up, be in the moment and be open to opportunities as they present themselves – not pennies. My mother was expressing a view of scarcity. Some inherited assumptions included these ideas:

- You have to fight for everything in life
- Don't trust anyone
- More is always better
- Work is the way to security because you need lots of money
- Save as much money as possible

Perhaps you agree that some of these are important, I no longer do. What I discovered is that these beliefs of scarcity, of living life in constant angst, fear and never enough were so ingrained in me that I could not examine them let alone try to change them. These internal messages in my head were preventing me from experiencing any long-lasting happiness or joy; in effect these beliefs amounted to self-sabotage.

Living in scarcity is protecting what you have. Those in scarcity believe the world is a zero-sum-game[44] which means for you to get more, someone has to get less; a world where you have to fight for everything and put on your armor before leaving the house.

Rather than cling to what you have, I found it better to strive for a life of abundance; opportunity is everywhere and you just need to notice it around you every day. There is enough to share with others and that it is okay to share your thoughts and feelings with trusted people. In this mode of thinking you are more likely to be generous, supportive and open to others without expecting a payback. With this view the universe provides for all who seek. I first became aware of this concept from the best-selling book "The Secret"[45]. As an engineer this is definitely a stretch for me since there is no objective evidence to support this idea. However, I find this way of thinking far better than the scarcity laden zero-sum game of economic theory which purports that I can only win if you lose.

An abundant universe raises all hopes

It's not an easy transition. It takes the will to want a better life; it requires effort, time and a willingness to let others help you. Do you want to make this change? If so, then follow these two simple steps:

1. Just <u>believe</u> that you are entitled to a better life – we all are!

2. Do it ... which is the rest of this book.

Credis Quod Habes

Years ago I saw a magic act at Caesar's Palace called: *Credis Quod Habes et Habes*. The entertainment was fun but the phrase sparked my imagination. I kept the ticket shown for over 20 years. Could it be true? Latin for "whatever we believe becomes reality"? The Matrix[46] plays with our concept of reality but the idea that we can create our own reality ... wow! And there are some intriguing scientific lines of investigation that are worth mentioning.

A fascinating experiment was conducted in 2012 to test this idea. The subjects were connected to an MRI machine that would display and record the active areas of their brains while they watched videos[47]. A little later while still connected to the equipment, the subjects were asked to recall their experiences. To the researchers' astonishment, their brains lit up the same way! There was no difference between the memory and the actual event. So how do we know what is real and what is a really good memory?

Ryan T. Howell is a Professor of Psychology at San Francisco State University and one of his research topics is the relationships among wealth, consumption and happiness[48]. He concluded that people enjoy their vacations more if they spent their money on activities rather than things. The memories were much more enjoyable and they can be revisited at any time. So why not assume we have the power to make our own reality – and make it the best we can imagine? It all starts in your head.

The brain we wake up to in the morning is not the brain that we put on our pillow at night – it changes all day long. Tell yourself positive affirmations every day to counter all the negative ones from the world. When you do an affirmation, keep it simple, clear and repeat it often. If you say it out loud, especially in front of others, it will become more tangible and ultimately manifest itself into your new reality.

If you are struggling to change your patterns, do ONE thing different. More than likely this will get your brain on a different track that will break the auto-pilot routine. For example, if you are out shopping for clothes, try one item that you have NEVER put on and see what happens. Notice how it makes you feel; it may inspire you to try something else, give you an idea for an afternoon activity or awaken a desire you had forgotten about.

You can do anything

Well, that's not quite true. For example, few people in the world can qualify for the Olympics pommel horse competition regardless of age and financial resources. However, the spirit of this idea is that for the vast majority of activities out there, you can participate. And if you discover at age 65 that your passion is truly pommel horse, you could be a coach, raise money for the team or even build pommel horses.

Having at this point done considerable work to turn my life in a new direction, I decided to learn swing dancing. I found a fun and friendly dance studio that offers a wide range of dance styles with motivational instructors; even the name of my favorite dance style is fun – Lindy "Hop". Some months later after getting competent in the dance style, I went to Downtown Disney where they had setup an outdoor dance floor complete with a 6 Piece Band. The floor was packed and surrounded by admiring park visitors. First time dancing to spectators – really cool because I remember being on the outside looking in just a few years ago. And then my night became magical. I was looking for my next dance partner, turned to my left and saw ... Serrah.

There are a handful of dramatic moments in one's life that elicit uncanny clarity and joy – that change everything. This was such a moment for me. And it was Valentine's Day. She returned an intrigued head tilt and said I looked familiar ... (to be continued).

Rainbows and Unicorns – Not!

Although I now spend an inordinate amount of time in a positive, happy space, my goal is not to eliminate negative emotions. In fact, I found out quite the opposite – it's critical to examine, explore and embrace negative feelings. As is a central Buddhist tenet[49], it's about embracing life in all its facets and to face the bad stuff head on.

Brene Brown has a great expression: "Lean into the discomfort"[50]. Yeah, I know – who wants to do that!? It took me some time to figure out what she was talking about. If you try to suppress or ignore negative feelings, two bad things happen. One, you're always fighting them and then you feel worse when you lose the fight, which is inevitable. And two, if you suppress the negative emotions, you also suppress the positive emotions because they are all part of the same system – you! This revelation was huge for me because up to that point in my life it seemed that I could not feel at all and consequently could only connect with people at a superficial level. There was a hollow emptiness in my soul that no amount of ice cream or playing solitaire could fill. And there's a potential third consequence; once you feel even worse about yourself, this is fertile ground to pick up an addiction, obsession or other serious disorder.

Priorities

Do you let the phone or emails interrupt you? Why? When this happens to me it's because I'm bored or restless and looking for an excuse to waste time or not deal with something that needs addressing. It is difficult to get a major task done or enjoy a moment when YOU let others or technology get in the way. Allocate a particular time to get in touch with telemarketers and others. And please don't tell me you have given your phone a higher priority than a real person! Remember, the real, ultimate goal of all human beings is to connect with another warm body. Brene has a phenomenal TED Talk that explains shame, vulnerability and our fundamental need for connection in a 20-minute talk that powerfully encapsulates her 7 years of research and 2 years of personal development that provided her clarity and really helped me on my journey[51].

You do what's important. Stop fooling yourself with: *"Yeah, I should lose some weight; I need to cut back on gambling; I need to make more time with the kids; I really should show my partner that she's important; I should write the last 40 pages of my book …"*. Oops that's me. Just do something about it or choose that you will not and live in peace. You will avoid feeling bad whenever you think of these unrealized and unplanned desires.

Decide right now!

My weight is okay and I don't really need those last 10 years of my life, or get yourself an exercise goal. As for me, I will write 10 pages on Sunday at 10am and have set a calendar event on my phone. (Pages were written at 4PM on the scheduled Sunday).

Don't know what you want? Here's a fun idea, every month do something you've never done before. It's a great way to meet new people, bring some excitement to your life and have some new stories to share with your friends. Can't wait to share my adventures when I learn beach volleyball next month! Check Meetup[52] groups or evening college classes in your area for a wide variety of activities. I've tried a number of both of these with good success. Another benefit is that other people find it really impressive that you are trying new things. You may even inspire them to do the same and join you.

Have you ever had to deal with an overwhelming situation that was a whirlwind of chaos and had no idea how to even begin tackling it? Here's how I deal with those extreme situations from my engineering background – start with a corner. It's a little like doing a 10,000-piece jigsaw puzzle where you can easily gather up all the straight edged pieces and put the border together; start with what you know. In the maelstrom before you, is there a piece that you recognize will fit with another? Great! Start there. Completely disregard the rest of it. Then focus on another couple of pieces. Eventually you'll be able to start putting the clumps together and then – snap! The big picture will come into focus and you'll see the total solution.

For example, after a multi-course dinner the pile of dishes in the kitchen can seem overwhelming. Well, I don't want to break the crystal wine glasses so I'll take care of those first: soap, rinse, dry and put away – done! Now if I stack all the plates together, I'll have room to work. I can now safely unload the dishwasher. Then load up the dirty dishes, gather the utensils. Now tackle the pots and pans. Need to look around at this point because I have a habit of not noticing stuff left on the stove. Then a quick wipe down of all counter tops. *Et voila - c'est fini!* I can now go to bed with serenity knowing that things will not be growing amid the sink clutter to be battled in the morning.

Building an Amazing Life

When I was playing with animation, I obtained an amazing reference volume called "The Illusion of Life"[53] by Frank Thomas and Ollie Johnston which chronicles the history of Disney animation – truly fascinating. The secrets of bringing those beloved characters from acetate to the screen and into our hearts is described in loving detail by two Disney animators who worked on such iconic features as: Pinocchio, Snow White and The Jungle Book. Any animation consists of a series of drawings or frames that progress in time at least 24 images every second to give the illusion of motion. Our eye-brain system cannot discern anything faster as distinct things so the series of drawings is interpreted as motion.

Life is a string of moments

In the same way, I believe our lives are a series of moments strung together to give the illusion of a grand plan from birth to death. The reality is that those discrete moments of RIGHT NOW are what is real. Put one, two or six such moments with your kids together and you have an hour. Add up 10 to 100 special moments with your family and friends results in a day that has been seized. And with this perspective you will not worry about how quickly a year has gone by and lamenting regrets because you will have experienced thousands of special moments that fill your life. So pay attention, enjoy and relish each and every one of these special moments because "there are no ordinary moments"[54].

Affirmations

It is inevitable that at times we will feel doubt, uncertainty and fear. The good news is that we have the power to stop these negative thoughts. A potent weapon to fight back is the use of affirmations. These are positive thoughts about what we want in our lives. Our goal is to displace the negative gunk with aspiring good stuff – positive thoughts attract positive results.

Some time ago I adopted my first affirmation and said it aloud every time I saw a sunset: *"I know the light, warmth, beauty and majesty of the sunset is a special gift from the Universe that will fill my hollowness. I will feel love and acceptance for who I am"*. The "Universe" is my higher power.

An astounding thing happened after reciting this affirmation for three months. One evening, that didn't seem different than any other, I said: *"I know the light, warmth, beauty and majesty of the sunset is a special gift from the Universe that fills my hollowness. I feel love and acceptance for who I am"*. The future tense suddenly changed to right now! Instead of saying "... that WILL FILL my hollowness" the words that came out of my mouth were "... that FILLS my hollowness". I was flabbergasted at this realization – I believed my own words!

A Special Sunset Affirmation

I added a sunrise affirmation as well. All I need is to see the sun at sunrise or sunset, which prompts me to immediately blurt out my affirmation; and every time, I feel a warm hug from the sun that makes me smile, sigh or even shed a tear of joy.

Now back to you. You did an Affirmation exercise in the "Decisions" section previously. How is that coming along? Are you saying your affirmations out loud? Do you believe them? Haven't done it yet? Now is a great time to do so.

Have you ever opened a fortune cookie with a perfect prediction of your future? Were you perhaps wishing for a different one? How about taking a strip of paper and writing your own fortune! Put it next to your other three affirmations.

Spin It Around

When I start feeling sorry for myself and spinning in my own head I often use a technique I call "Spin it around" – how would I react if the current dark thought in my head came from one of my friends? For example, I began to feel sad when struggling with finding that special person after so many failed attempts. The dark thought that began forming in my head was: *"nobody is interested in me"*. Using "Spin it around" my thought is framed as though a dear friend was saying these words in all earnestness. Without hesitation, I would see the truth behind the fear and gently remind my "friend" that he had politely declined pursuing "I am interested" signals from three ladies and passed over 30 online dating matches last week alone. Women are interested in you and there is simply not a match; you just haven't found her – yet.

It's not about white-washing your thoughts and eliminating all negative ones like using bug spray for your emotions. Rather, it is figuring out how you are feeling, acknowledging honestly why, putting it into perspective, accepting it then moving on. The process invariably shrinks the dark cloud in my head back to its initial miniscule size once exposed to light and the feeling passes much faster. The struggle no longer persists for days which could accumulate other negative thoughts into a maelstrom of negativity.

It seemed easier for me to be kind to a friend rather than with myself. My dysfunctional family behaviors taught me to be self-sacrificing and to neglect my needs, which may initially sound noble but are really not. Such an attitude invariably leads to isolation and resentment which helps no one. I deserve to treat myself at least as well as I treat my friends – and now I do.

Brainstorming

When I was in grade 7, our history teacher Mr. Nesbitt, did not make it to school and so under short notice the Vice Principal stepped in to fill the hour with 30 rambunctious pre-teens. He was not familiar with the lesson plan nor, I suspect a big fan of history, like the rest of us in the class. What he showed us was amazing and became a tool that I've used regularly in my career. He demonstrated how to brainstorm.

It is a novel approach to finding a solution when you have absolutely no idea how to proceed. It's a bit like throwing a plate of spaghetti at a wall, see what the pattern looks like then pick off the piece that's the perfect solution. Here's how it works.

The problem he setup was a long distance bus ride in which you are seated next to an elderly woman who is blowing bubblegum bubbles and it is driving you crazy. There are no other available seats and you are not allowed to stand or get off the bus. You have to deal with the problem. The class was not impressed and some were even thinking the history lesson would be better. However, in short order the VP explained that collectively we were going to brainstorm a solution. You could propose ANYTHING without judgement of any idea. The students began to contribute: *"put tissues in your ears", "turn on a radio", "hum a tune", "grab the gum"*. Then someone objected: *"Ooh that would be sticky!"* The VP re-iterated the no judging rule and other students took on the policing duty as the suggestions began in earnest: *"tell her to stop or you will pop it in her face", "poke her in the ribs", "throw her out the window"*. Yes, I know, the poor old lady. But she was purely theoretical and the kids were learning teamwork. Within a few minutes, the blackboard was covered with over 50 suggestions.

Now for the second phase of the process; acting as the moderator, the VP began to review all the suggestions with the class. A discussion ensued to eliminate extreme suggestions that were impractical if not outright illegal. Following a process of consensus, merging of ideas and editing a final solution was crafted: *"Ask the woman to refrain from gum chewing because it was becoming distressing"*.

The key to success is to write down every single idea no matter how crazy, absurd or unlikely it may be. This process is using your creative right brain which we will discuss in more detail in a later chapter. The right hemisphere does not do well with criticism so there must be NO JUDGEMENT in this step, especially if you've asked others to help you. You are collecting ideas from all sources, inspirations and whims. One crazy thought may lead to a bizarre suggestion, but when you turn it sideways it may reveal a nugget that becomes the keystone to the solution.

EXERCISE – Brainstorming (60 min)

Now it's your turn. For this exercise you will need help from at least two people.

1. Think of a problem that you have been struggling with.
2. Ask a few people to help you and set a brainstorm date.
3. Be sure to explain the two rules: any ideas are worthy and there are no negative comments about any suggestions.
4. Collect all ideas on a whiteboard, flipchart, poster board or other large media. You need at least 25 – think outside conventional solutions.
5. Thoughtfully review the suggestions as a team. Merge similar solutions, modify or eliminate unworkable ones until you can circle the best final answer. In no time you will all feel like you're back in junior high!

Give it a Rest!

Sometimes all the meditation, concentration and brainstorming just doesn't get you results. So give it a rest – literally forget about it. Your conscious mind knows the problem and all the parameters really well. Let your sub-conscious do its magic. In the morning or in the middle of dinner or (this happens to me) in the shower – bam! The solution appears completely unbidden. And I just say: "Thank you Universe". The more this happens the more you will let go of forcing an answer on your timetable and accept that the answer shall be delivered.

While writing this book I would have little snippets of ideas for chapters at unusual times, like on the treadmill at the gym. And I would look around helplessly as the inspiration faded with neither a pen or whiteboard insight. I purchased a small flip pad that fits in my pocket with a small pen, I was now ready to jot gems at a moment's notice!

Sometimes these flashes of brilliance didn't fit anywhere nor did they make a whole lot of sense. But I stuck them in the "Other Thoughts" section at the end of the manuscript. Then when I was staring at a blank screen, I would grab one of these idea sparks to get me rolling on pumping out some prose.

Let it go!

Scientific Process

Inside the Mind of a Scientist

Have you ever wondered what it's like being a scientist on the trail of a Nobel Prize? Okay, probably not but indulge me for a few paragraphs. The essence of a scientist at their core is an innate sense of wonder and curiosity about the world around them. You may see where I'm going with this – curiosity about one's life.

Albert Einstein as a young adult sitting bored in a Swiss patent office in 1905 wondered what it would be like to sit on a photon (a tiny, tiny sub-atomic particle that transmits light) moving at exactly the speed of light[55]. From that creative thought sprung his famous theories of relativity which completely changed his life and everybody else's too.

The Scientific Method

The cornerstone of modern science is the Scientific Method. It is a way of exploring our world to learn how it truly works without bias. It involves 4 basic steps: ask a question, collect lots of data, propose a hypothesis then test it[56].

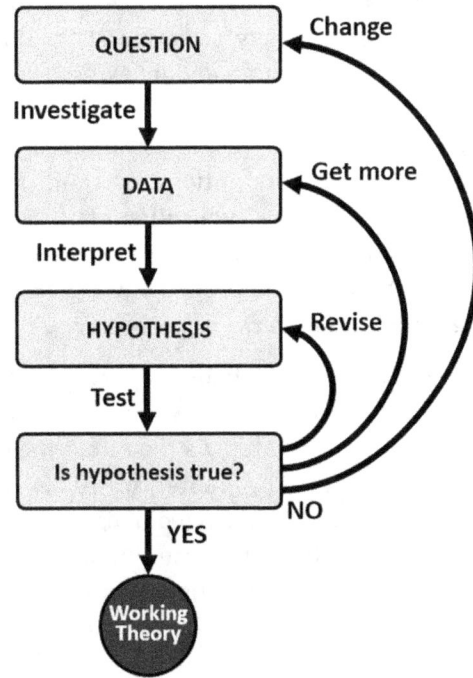

Scientific Method Flowchart

The scientific method begins with formulating a question about the world - what you want to better understand. Then you investigate the topic by collecting objective data through experiment and observation. It's vital here that your data maintains the highest integrity with accurate, un-biased observations. The next step is to interpret your findings and propose a hypothesis that explains your data. This is the fun part because you get to figure out the riddle that nature has provided. The last step is the critical test of your hypothesis. The key thing here is to make a valid prediction as a result of your hypothesis. For example, if your hypothesis is that gravity is the bending of space-time in the presence of large bodies, find out if starlight is altered by our sun and could we observe this shift during an eclipse? It does, as Sir Arthur Eddington successfully performed in the first experimental test on Einstein's General Theory of Relativity on May 29th 1919[57].

The scientific process is an iterative process; if at the end it does not answer the original question you can revise the theory, get more data or change the question. The question is key because questions are worth so much more than answers – they are the doorway to knowledge, experience and the fun of discovery! The entire foundation of quantum mechanics and our knowledge of relativity and gravity is the result of Einstein's question about sitting on a light particle.

The Scientific Method has endured since the 17th century because it is simple, independent of any individuals, groups, agendas or assumptions and it is self-correcting. Any scientific claim must be backed up by publishing the collected data and duplicated by at least one other team (preferably on the other side of the globe). If the results cannot be replicated, the claim is dismissed. Some ideas persist for a long time without verification like the notion that outer space was filled with a substance called "the ether" which people thought was necessary for light to travel from the sun. The concept was seriously considered for over 200 years because the technology did not exist to measure the speed of light accurately so the hypothesis could not be tested.

Michaelson & Morley in 1887 figured a clever way to test the concept by bouncing light between two mirrors and measuring its speed in different directions[58]. If light propagated via the ether, the observed speed in the direction of motion should be greater (speed of light plus speed of ether) than in the opposite direction (speed of light minus speed of ether). But regardless of the direction of measurement, the speed was always the same. The result was completely unexpected and provided a catapult for future discoveries.

Michaelson & Morley were able to test the ether hypothesis because their experiment allowed an exceedingly accurate measurement of the speed of light – a whopping 186,350 miles PER SECOND! To illustrate this fantastic speed, image you are in your backyard tonight. Aim a laser pointer at the moon and briefly turn it on. It would take 2.5 seconds for that light to reach the moon and bounce back to you. Now if you did the same thing the next morning with the rising sun while sharing a cup of coffee with your loved one, it would take about 16 minutes to reach it and return. And since we're talking astronomy, if you launched a light beam at the nearest star, the Alpha Centauri tri-star system, then decided to add to your family after breakfast, the light would return when your son was in the fourth grade. Incidentally, a planet similar in size to the Earth was found circling in the habitable zone of Proxima Centauri in 2016[59].

If a scientist is shunned by his peers because his discovery is too difficult for them to accept, or there is too much ego or money invested in the status quo (yes, scientist have foibles too) in time the scientific method, the data, will prevail. Gregor Mendel[60] was an Austrian monk and radical scientist who grew large gardens to support his fascination and experiments hybridizing peas. His idea that plants could change over generations in a predictable way was heresy in 1866 when he published his findings[61]. But 36 years later in 1902, William Bateson revived his ideas and Mendel was hailed as the grandfather of genetics by a more receptive audience.

The Greek philosopher Aristotle was one of the first to examine the world in a methodical way around 350 BCE[62]. Among his many ideas was the concept that the speed of falling bodies depends on their "earthy" quality or weight. This idea was unchallenged for almost 2,000 years until Galileo Galilei asked a question: *"Do heavy objects really fall faster than light ones?"* He didn't think so and in 1589 he setup an experiment to find out[63].

Falling Objects Flowchart – Version 1

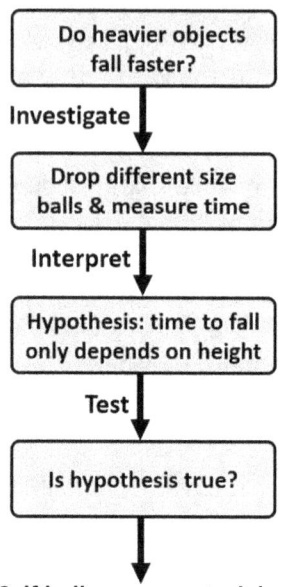

YES, if balls same material
NO, if balls different materials

Galileo famously dropped different sized steel balls from the leaning tower of Pisa and measured the time for them to drop. He collected data so as to test his theory that the time to drop depended only on the height. So is it true? The cannon ball landed before the musket ball – but only by a few inches. Galileo was convinced that he was on the right track so did more experiments with different kind of objects and noticed that the shape of the objects impacted the time – air resistance! Galileo unexpectedly discovered the concept of terminal velocity which is why parachutes work.

Falling Objects Flowchart – Version 2

He modified the question of his experiment to: *"Do heavier objects fall faster IN A VACUUM?"* One hundred years later the vacuum pump was invented which verified that a feather and a steel ball do indeed land AT EXACTLY THE SAME TIME. In 1971 astronaut David Scott on the Apollo 15 mission dropped a hammer and a falcon feather (the lunar module was called Falcon) on the moon (there's no air so no air resistance) and they too hit the lunar surface simultaneously[64]. Galileo more precisely derived the equation that could be used to predict the time it would take a body to fall from a specific height. The equation is shown on the next page where "t" is the time to fall in seconds, "h" is the height of the drop point in feet, and "g" is the earth gravitational constant which is approximately 32 feet per second squared ("g" varies a little depending on altitude i.e. distance from object to the center of the earth).

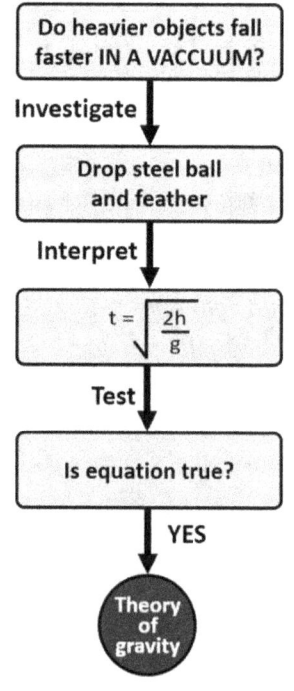

Ball Drop Equation and Table

t (seconds)	$h = \frac{1}{2} g t^2$ (feet)
1	16
2	64
3	144
4	256
5	400

If you're not especially fond of equations, look at the table showing how far it falls during a 5 second journey. If you dropped an object in a vacuum, of any size, shape or weight from a high tower, it would fall 16 feet in the first second, 64 feet by the next second and 400 feet from where it was released in only 5 seconds.

Galileo's question and subsequent discovery allowed big leaps of progress in astronomy, physics and engineering. So ask questions and follow where they lead because they will enable you to engineer a new life for yourself!

Occam's Razor

There is a scientific principle called Occam's Razor that helps scientists when confronted with multiple hypotheses. William of Ockham (1287 – 1347) formulated the idea that in the event of multiple explanations, the simplest or the one with the fewest assumptions is usually the correct one[65]. For example, let's look at two different explanations for gravity:

1. Gravity is a force exerted between any two large objects that attract toward each other due to the distortion of the space-time they occupy.

2. Gravity is the handiwork of invisible pixies that permeate the world and they ensure that all things fall in the exact same way every time for everything on the planet.

Until we developed rockets and saw astronauts floating in their capsule beyond the influence of the earth pulling them down, there were many who believed the second definition or a similar variation. Although there may be a handful of holdouts that would say: *"Everyone knows pixies can't live in outer space; it's too cold there for them."*

Occam's razor can be helpful in everyday life situations. You may think that one of your peers at the office has stolen your Oreo cookies from your desk drawer while you were away at a meeting. Or you simply ate them yesterday and forgot to get more. Which is the likely solution? Look to the simplest one.

Personal Progress Method

We can apply the Scientific Method to advancing our personal progress with similar success. The Personal Progress Method Flowchart is shown below. It also begins by posing a question. Your problem may be for example "I don't have enough time". Posed as a question, this could be:

- Why don't I have more time?
- Who is taking up all my time?
- Where am I wasting time?
- How can I get more time to do the things I enjoy?
- How can I spend more time with my significant other?

Finding the right question is essential for success so spend some time figuring this out. In our example, the last one is probably the best one because it is framed positively, specific and meaningful.

Personal Progress Method Flowchart

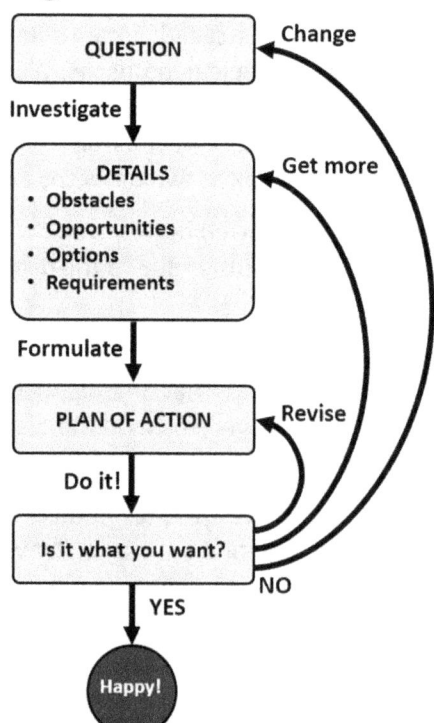

In the next box investigate the details surrounding your question: obstacles, opportunities, options and requirements. In our example, a good start is to find out where you're actually spending your time. Start a diary and jot down your activities and time spend on each to the nearest half hour. You may want to add a rating on how much you enjoy the activity (e.g. A, B, C or required). Do this for a week then tabulate the results something like this:

ACTIVITY	TIME	RATING
Sleeping	55Hrs	required
Eating	17Hrs	required
Dining	3 Hrs	A
Work & Commute	45 Hrs	required
TV	15 Hrs	C
Email & Internet	12 Hrs	B
Household Chores	8 Hrs	required
Grooming	8 Hrs	required
Shopping	5 Hrs	A
TOTAL		**168 Hrs**

Once you know where your time is going you can decide if you want to reallocate your priorities differently or just keep doing what you're doing. This is the next box, to formulate a plan of action. In our example, there may be some surprises in your summary. Watching TV is 15 hours and yet has only a C rating. Maybe there's another activity, perhaps something new that you could do with your significant other that is equally relaxing to your customary two hours of TV after dinner. Maybe going out to dinner more often is more desirable than all that time on Facebook with your friends. Defining a category "Quality Time with Significant Other" on the spreadsheet would also help focus your priority.

Now do it! Talk to your significant other about your plan, your commitment to spend more time together, or surprise them at a special dinner to show that you mean business. After a few weeks, ask your original question and see if your solution is working. If it isn't, don't give up, make some tweaks to your plan. If you missed something important in the investigative part, collect more data and pull out your diary for another week. Yes, it takes work and time – this is required if you really want to Engineer a New Life.

Be aware and make deliberate choices; don't just live by default. If your question was *"Why don't I have enough time to become a café musician?"* and after analyzing your time budget you decide not to pursue the musician lifestyle, you will feel better knowing you are spending your time exactly the way you want. And you can let go of other options. Converting all of the "shoulds", "woulds" and "ifs" to I AM or I AM NOT will make for a much happier life.

EXERCISE – Plan for #1 Life Goal (60 min)

Now it's your turn! Go back to the second exercise in this book and rank your Life Goals from 1 to 7. Copy your number one goal below.

Fill in the rest of the Personal Progress Method flow chart below to turn your desire into reality using the previous example as a guideline. Start with the central question that is hindering you from moving forward with your goal. Write it down in the first box at the top of the flowchart. If you have multiple questions regarding your number one life goal, write them all down and tackle the most difficult one. You can repeat the process for the next one after you solve the first one.

Determine what data would be useful for you to consider in answering this question. What's stopping you? Put these in the "Obstacles" section. If you need more room use the white space to the left of the "Data" box. Do the same for "Opportunities" that is the things, people or situations that may help you answer your question. "Options" are different aspects of the topic that may give you a fresh perspective on your question. This is brainstorming time so jot down anything that comes to mind. And finally "Requirements" will give you clarity on the essential things that pertain to this issue; sometimes focusing on the basics can be useful. You may need some work and time to get all the data you need from this step, like the diary used in the example.

Once you have collected all the data, you can begin to find some patterns, some cause and effect connections. Be patient, you may need to sleep on it for a few nights and let your sub-conscious help you out. From this insight, come up with a plan to take action.

After you've executed your plan check to see if it's what you want. If yes, then you're done. If not, then you may need to revise your plan, get more data to figure what's really going on or perhaps change the question and rethink the rest of the process.

My #1 Life Goal:

_____

_____

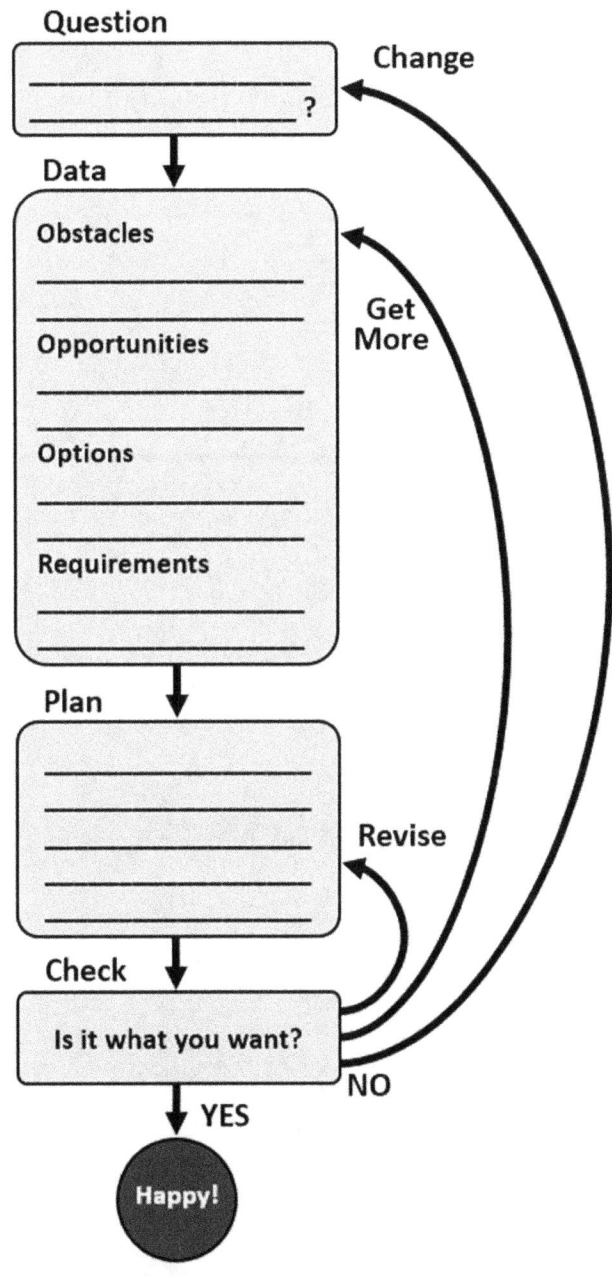

Say Hello to your Right Brain

Life is so amazing! I'm on a plane (was up at 4:45AM) with two connections and 6 more hours of flying ahead of me. But I have my magic headphones on, bopping to Avril Lavigne and so grateful for the beautiful sunset at the beach last night, the fun dancing on Saturday and reconnecting with a friend. It is all about your state of mind – which you can completely control.

For the last few days I have been outside my own head paying attention to the world around me and reveling in it! I'm not being a victim, not complaining, not waiting for an event – just enjoying ... right now. And you can do it too – whenever you want. Believe it or not, I think the headphones are a big part of this moment for me. Music, like most forms of art put you in right brain mode (the creative side) which is the domain of now; the left brain is in charge of past, future and pretty much everything else that is quantifiable because it knows the other half is pretty useless for these tasks. Actually the right brain has no idea what a "Task" is and it has no interest in learning.

EXERCISE – Soul Smiling Quotient (15 min)

Try this simple exercise right now to experience what I'm talking about. On a scale of 1 to 7, rate your "Soul Smiling Quotient" that is to say how you feel right now. Score a one if you think this is a lame exercise but will grudgingly do it anyways. You get a seven if the last sentence made you laugh out loud. It's not that hard, please do it now.

Now turn on one of your favorite songs and engage with it – sing along, pick up your air guitar or practice your 8 count swing footwork, even an impromptu aerobic routine in your underwear will work – just do something! And yes, right now please.

Did you do the exercise? If not, go back and do now.

Seriously. Do the exercise!"

Welcome back. It was fun, right? Now honestly rate your Soul Smiling Quotient again. Most people score two points higher. (Feel free to rate yourself an 8 or 9 if you were already in a right state of mind before you played the music. But not a 10; the right brain is most certainly not about perfection. And that's just by listening to one song! Imagine how you would feel if you sang for 30 minutes, or doodled a picture or went to a concert.

Geckos

I was inspired recently while working on my book to create a short animation of a gecko; something goofy to make people smile. So I started sketching geckos like the one below.

I was struggling with getting perspective angles and had the idea to create a gecko model. So I bought some clay (you can get any color you can imagine), had a blast and made this little guy below – he likes to hang out in the garden.

It's that easy! Just get yourself into creative mode by listening to some tunes then follow your inspiration to have a wonderful time in right brain mode. When you get back to your logical half you may find that the creative break has solved a nagging problem or given your perspective on an issue that allows you to see that it isn't a big deal after all.

Get in touch with the Universe

Jill Bolte Taylor is a neuroanatomist and an amazing woman who opens her spellbinding TED Talk with a real human brain in her hands. She had the strength, courage and scientific dedication to take mental notes while she experienced a massive stroke in 1996[66]. Over the course of four hours, while her left brain (the practical half) systematically shut down, she discovered the hidden power, beauty and perception of the right hemisphere. For most people, daily thoughts are dominated by the left analytical brain. She describes an overwhelming feeling of peace, serenity and connection to the world around her. After her seven-year recovery, she dedicated her life to getting back to that feeling and enlightening others so the world could be a better place for everyone.

The bottom line is that our right brain works fundamentally differently than our left side. You can't survive living only in your right brain – you would never get anything done (including making money, eating and getting out of bed). You can however survive living only in your left brain – I did for many years. That's the realm of responsibility, solutions, work ethic, sense of self, past memories, future planning and "To Do" Lists. Yes, you too can be a permanent resident of the left side with regrets and shame of the past, anxiety and fears of the future, struggle for perfection, never quite meeting expectations, judgment and addiction.

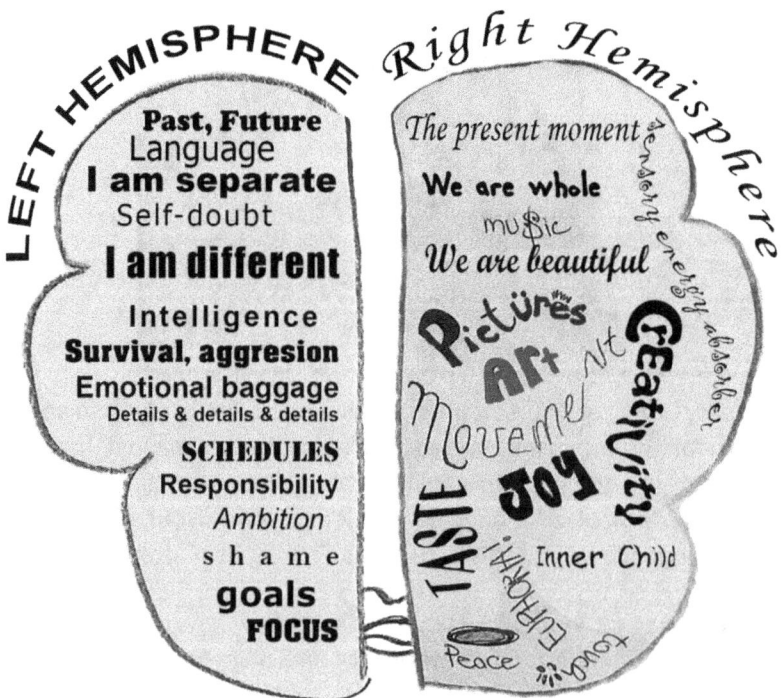

Left and Right Brain Functions

The right brain is all about beauty, joy, peace, serenity the present moment – right now and connection to the energy of everything around us. Sound nice? You can't survive living only in your right brain – but wouldn't you like to visit more often?

Let's do that right now! Yes, pun most definitely intended.

EXERCISE - Visiting your Right Brain (45 min)

You will need a quiet space with table and chair, large pad or sheet of paper (8.5" x 11" will work nicely), working pen and a timer set for 10 minutes. Go get them now.

Ready? No interruptions for 10 minutes? Great! Start the timer.

Get in a seated position with the pad before you. Now move the pad a little to the side (right side of the table if you're right-handed or to the left if you are left-handed). Grasp your pen in your writing hand and position in the middle of the pad. Now place your other hand on the table in a comfortable position where you can clearly see it. Look at your hand carefully and deliberately.

Now without looking at the paper nor lifting the pen off the panel begin drawing what you see. Start anywhere, and let the pen follow what your eyes see - each line, curve and crease. Do your whole hand or just the intricate detail of your thumbprint, anything will do. The drawing that you create does not matter. What is important is that your focus only on your hand for ten minutes. Let it become your whole world; nothing else is of any concern. Just enjoy the freedom of drawing what you see ...

How it works

In order to get access to your right hemisphere the left side has to relinquish its control. Not an easy thing since it likes to be in charge. Here are some typical left brain background thoughts that you may be familiar with: *"Did I do that right? Is it time for dinner? How am I going to pay the bills next month? What am I going to wear tomorrow?"* Many of these questions are important and it's a good thing our left brains track all this stuff. But sometimes it needs a break – YOU need a break from the barrage! The trick is to do something that your left brain finds boring – like drawing. Then the strong-willed left hemisphere goes away leaving the right half in charge.

The diagram on the left shown on the next page was drawn with the left brain in charge: *"Draw a hand, okay. I know what a hand is, five fingers, palm and wrist. Done. That required 13 seconds. And I don't see the point of any of this."* On the other hand, the right brain knows nothing about time, hands, fingers or completing tasks. Instead, the right brain is content to just enjoy the experience: *"Oh wow what's that? How cool, we get to draw! There's a lovely slender blob. Sinuous line that bends over here ... and then over there. Then it crosses over this bumpy thing then goes on and on and ..."* Bing! Whenever we give control to the right brain we definitely need a timer otherwise nothing would get done.

Drawings by Left (analytical) and Right (creative) Brains

We live in a dominantly left brain world. What you know is more valued than what you feel. Quarterly bottom line results are what drive most businesses, while a beautiful inspirational picture or sculpture is scorned or ignored.

Most people are similarly left brain dominant. Because our brains are cross-wired, the right side of our bodies is controlled by the left brain and vise-versa. There is a clever saying for south paws: *"If the left brain controls the right side, then only left-handed people are in the right minds"*. And there is a correlation between handedness and specialized brain hemisphere functions like language, art and music[67]. If you are in the majority, people who struggle with creative endeavors here are some ways to access your right brain.

Meditation

Meditation is another way to access your creative right side. There are many ways to meditate which have varied emphasis on the following elements: controlled breathing, posture, sequentially relaxing muscles groups, having a point of focus and clearing your head of any thoughts and the famous "OM" – the sound of the universe[68]. Whatever method you use to get there, you are essentially visiting your right hemisphere to embrace the non-verbal, non-judgmental, in the moment part of your brain. Here are a few other options that work for me.

To calm my thoughts, I like to try to visualize – nothing. I imagine non-existence. A black hole. When my thoughts start to formulate ideas or pictures and deliberately return to a blank slate. It is surprisingly difficult to do. I use this technique when having difficulties falling asleep as well.

If you like the total silence option, you have got to check out noise-cancelling headphones! They contain microphones near each ear to detect the sound around you then electronically add the exact opposite waveform so that your ear hears almost none of it. I recently purchased a pair and took them on a fully packed flight. As soon as the engines started revving up and the blaring instructions began, I slipped on my headphones, flicked the switch and was immersed in a Buddhist-like chamber with gentle murmurings in the distance –something about releasing a buckle. Then I plugged in my tunes and rocked-on for the entire four hours! I even was bopping down the aisle with headphones on my way to the restroom. Without a doubt, I was the happiest traveler on the plane. And when we landed, I was rested and vaulted down the concourse to the rental car ready for the next part of my business trip.

Whatever method you try, remember that you don't need any preparation or scheduling – just do it. Give yourself the rejuvenating gift of meditation. How about right now?

EXERCISE – First Meditation (15 min)

Set your timer for 15 minutes. Go to a quiet place, get in a comfortable seated position (preferably sitting with your back snuggly again a backrest). Take nice long breaths in, nice long breaths out and either with your eyes closed or fixed on something interesting to you like a flower – just be there in the moment. No agenda, no worries, just you and the flower – the only things that exist in the universe right now.

If your timer goes off and it seems like it was only a moment ago that you set it, you are doing it! If it was torture to just sit there, tomorrow try a different method, location or time of day. And feel free to give yourself another 15 or 30 minutes if you want to go back to your sanctuary.

Learn to Play

Playing and passionate activities (of all kinds) have big health benefits too. According to some medical professionals, the perks of play for adults include: relieve stress, improve brain function, stimulate creativity, keep you feeling young and improve your connection to others[69]. The family that plays together, stays together.

Power of Music

Music stimulates the right hemisphere, overrides the critical left brain and has highly therapeutic properties. You can breeze through boring tasks while listening to music because you are tapping into the part of your cerebral cortex that is in the moment and has no idea about time. And you just feel better - you can't be depressed while singing or playing an instrument. No doubt you have seen toe tapping, head bopping, air guitar strumming and big smiles on people who are engaged with music – be one of us! Consider it as a regular fun activity. Even singing in the privacy of your car can help ease tension while in traffic.

Music is a big part of my life now: playing piano, dancing to it and listening instead of other alternatives like TV, videos and newscasts. I think this change has really allowed me to spend a lot more time with my right brain engaged and hence more connected to the present moment. I suspect it's a big factor as to why I'm happy and joyful so darn much of the time. Add music to your life – lots of it! It can lift your spirits, take you out of your head and sometimes even give you a fresh perspective.

Delight of Dancing

From first-hand experience, dancing is a phenomenal activity! It's fun, there's music, you meet people, it most definitely is a right brain activity – and you even get some exercise as a bonus! There is a dance term called "connection" which refers to the non-verbal communication between two people dancing together (the leader and the follower). Because of the music, it is impractical to call out to your partner "we're going to do move 526 now". The areas where your bodies make contact (hand to hand, hand to shoulder, hand to waist) establishes the channel of communication and the location/degree of pressure is a cue from the leader to the follower of what is happening next. Experienced dancers know this communication or "lead" is a suggestion that the follower may accept or counter with a different move within the established framework. This is when the beauty of the dance appears – communication on the dance floor – just like relationships! No wonder dancing is so popular with the ladies – and they get to spin a lot too!

There is a wide range of dance styles to choose from: West Coast Swing, Foxtrot, Quickstep, Salsa, Tango, Nightclub Two Step, Waltz, Blues, Samba, Balboa, Samba, Rumba, Country line dance. If you have someone you want to dance with, find out the style they like best. It's been my experience that a happy dance partner makes all the difference to the experience. Don't have a partner yet? No problem, most studios accommodate singles by rotating every few minutes so you can meet many people quickly.

Try social dancing - it's fun, you can get up close with the opposite sex and you will spend some time in your right brain!

EXERCISE – Art Goal Review (15 min)

Revisit your Life Wheel Goals and review your "Art" goals. I highly recommend that you have at least one. Come on, art and music are fun. You deserve a little fun – it's an easy but important goal.

Calories In, Calories Out

"(Groan!) Yeah, I should have more fitness in my life". If this is your reaction, I can relate – it used to be mine too. Now fitness is part of my joy; actually it is the engine that drives my new perspective of life and propels me through the day. Sure, I used to exercise ... a little. It was always grudgingly, and getting my butt to the gym was so difficult. What changed?

Enjoy the Moment

My attitude changed. Getting fit now is a natural consequence of living in the moment for me. I used to focus so much on the past: *"I should have worked out yesterday - I feel bad - I need a milkshake!"* Or I would fixate on the future: *"I should definitely walk to the store tomorrow but ... probably won't have time so I'll drive instead"*. Now by living in the present moment, it feels so good to do a luxurious stretch on the beach before surfing, to slip into my cool workout clothes and run to the gym or feel the muscle burn doing bicep curls. I ENJOY EVERY SINGLE MOMENT.

Designed to Move

The human body is designed to move, walk, chase, forage, run, play and dance. Humans have been doing all these activities for over 100,000 years[70] with essentially the same body structure. It has only been in the last few hundred years that our ever increasing sedentary culture has reduced the opportunities to move the way we used to; big cities, cars, computers, grocery stores, cell phones all conspire to dramatically cut down the amount of exercise our bodies experience. However, our bodies still need and even yearn to be active. It's our minds that have become complacent with all the conveniences. If you think about it, it's not really a surprise that there's going to be problems; if you don't use your body the way it is designed to work, unused muscles atrophy, arteries clog with unhealthy cholesterol and oxygen starved blood slows down brain processes. With our altered programming I believe we have lost touch with what our bodies are trying to tell us: "I need to move!"

Typical Adult Male Life – Then and Now

Average Male Daily Stats 10,000 years ago

Weight: 140 lbs (lean)
Diet: 1,500 cal (on a good day)
Exercise: 2,000 cal
Foraging: sunrise – sunset
Skills: spear, bow, axe, tracking, tanning, fire starting
Activities:
 - Looking for food
 - Running from mountain lions
 - Climbing trees to avoid bears
 - Dancing around fire with woman

SAME BODY

Average Male Daily Stats Today

Weight: 210 lbs (not so lean)
Diet: 3,000 cal
Exercise: 1,000 cal (on a good day)
Foraging: 60 min
Skills: Word, Excel, order takeout TV remote, calling AAA
Activities:
 - Sit in front of computer
 - Sit in traffic
 - Sit at restaurant
 - Sit in front of TV with woman

One of the ways our bodies can truly rejoice in motion is to dance. I have been taking dance lessons for over a year and attribute some of my transformation to this wonderful activity. Dancing burns calories, reduces stress, fends off brain disorders like Alzheimer's[71] and increases the serotonin level which literally makes you feel better; no matter what my mental state arriving at the dance studio, I always feel better when I leave. And there even are more benefits!

From my personal experience, dancing builds confidence, you get to move with a partner (fabulous singles activity at any age) and something magical can happen. I had been enjoying learning new steps, making my dance partners smile (a really good sign you're doing it right) and moving to fun music.

It's not too late

With regular physical activity you can reverse the effects of a sedentary lifestyle and you will start to feel better about yourself – almost instantly! Be sure to start with small steps to avoid excessive sore muscles (a little discomfort lets you know it's working) or getting discouraged. You want fitness to be a regular, natural part of your life not something you check off on Mondays, Wednesdays and Fridays. By thinking this way, I am open to and embrace opportunities to burn a few more calories in my day like using the stairs or taking a small detour through the park if I'm early for an appointment.

I know that I will have extra energy, confidence and a sense of well-being all day long if I get in some fitness. That thought alone is enough to keep me on track because it is more important than checking everything off my "To Do" list or obsessing to get that work project perfect (95% will do just fine). Occasionally I will get a nice compliment from a co-worker while sprinting up the stairs – bonus! And all of this only motives me to do it all again tomorrow. I have noticed that it is so much easier to eat healthy after working out because I don't want to undo all my hard work.

Start Today!

So how can you start getting more fitness in your life? First, decide that you want it. If you're already thinking: *"Oh, that would be nice but ... I don't have time ... the right clothes ... the weather is not good this month ..."* then say OUT LOUD: *"I do not care about being healthy"*. How does that sound? You no doubt know there's a pile of logical reasons to be fit (live longer, fewer illnesses, faster recovery from injuries, more alert, burn more calories, to name a few) but the decision is an emotional one. And that is your choice. I only advocate that you be honest and make a decision – right now! If you choose to add physical activity to your day, start today. Here are some ideas that helped me:

> **SMALL STEPS** – Instead of committing to 90 minute workouts five times a week, start on a modest schedule that you can build on (e.g. 30 minute brisk walk twice a week). As you start to feel better, add to the frequency and/or duration depending on what works best for you.
>
> **FUN STUFF** – Choose activities you like or have always wanted to try. (Beach Volleyball is a recent addition of mine – fun!)
>
> **COMBINE** – Make time by combining activities like only catching up on the news while you're on a treadmill or cycling to some local errands on the weekend.
>
> **DE-STRESS** – Do you get stressed out? Perfect opportunity for some exercise! The next time you're irritated about a job situation, can't sleep, or worrying about the kids/relatives take a short run around the block or do some push-ups. Not only will the physical activity take your mind off your worries, the exhaustion will help you relax.
>
> **CARDIO** – Sustained brisk movement is key to work your heart and increase blood flow. My goal is at least 20 minutes of cardio vascular exercise 3 times a week. If you find walking, jogging or using hamster wheel-like machines at the gym a little boring, get some tunes! I did and wow what a difference! I now routinely stay on a little longer just to hear the end of the song.

I hope that you choose to add some fitness to your day. I believe it is essential to a happy and joyful life.

Diet

Diet can be a controversial topic with many opinions out there. I have no nutritionist or medical background so please consult with a suitable expert before making any radical changes to your diet.

Every day, every meal, every bite you have a choice. Do you want that tasty morsel or do you want to be a little slimmer and feel your clothes fit a little looser? I've actually picked up a delightful morsel, asked myself that question and then put it down. And you know - I felt really good about myself for it!

I have no idea how to lose 50 pounds in 50 days or the best micro-nutrients you should take. What I do know is how to keep my weight in a range that I am happy with and can share those experiences with you. From an engineer's perspective, there are some simple facts that allowed me to cut through the jungle diets, antioxidants and meal magic.

A human body is a living system that ingests energy in the form of food and converts it to what your body needs to keep the cells alive and muscles rebuilt after use. Any excess energy is stored for later use – the fat reserves. Our bodies have a long history of roaming the countryside in search of berries, nuts, grains and animal protein. Getting enough food meant survival so this was a top priority on our daily "Must Do" list right up there with evading predators.

Fat is a rare and coveted thing in the wild and when encountered is ravenously consumed. You may have noticed this response when someone brings in a box of Crispy Creams into the office morning meeting. We no longer need this programming in our lives but it is still there and hard to overwrite. Compounding this is the fact that obtaining food high in calories is as easy as opening the door to your local grocery store or driving to your favorite fast food establishment and speaking to a box. The natural controls to burn up the calories have also dramatically changed; I can't remember the last time that I was chased by a mountain lion or had to walk for days in search of a herd of deer. Unfortunately, our bodies and brains still behave the same.

Magic Calorie Equation

Engineers love equations. They elegantly cover many situations and are useful to predict future outcomes. Here is a simple equation for understanding food and exercise:

> **Calories Eaten – Exercise = Weight Gain (Loss)**

Here's how it works. Let's say you have a robust diet, 3000 calories a day, exercise often and with enthusiasm, in fact you work off all the calories you have eaten, 3000 calories. Then the math says 3000 - 3000 = 0 so there is no weight gain. If on the other hand you eat a similarly large amount but your idea of fitness is walking back to the buffet line, your will have excess calories that contribute to you weight – every day. (Assuming you keep this imbalance of energy). On the other hand, if your calories eaten are much less and you do little exercise, you will maintain your body weight. But if you eat little and do a lot of physical exertions, you will end up in the hospital because you're not providing your body with enough energy.

For those that eat smart and exercise a modest amount the equation predicts you will feel really good about yourself. If you eat smart and do an intense amount of physical activity you could be in the best condition of your life. See how cool equations are?

If you don't want to exercise you have two choices: eat less to lose some weight or eat with abandon and get bigger clothes. But either way, do not complain about it: *"I wish I was thinner. It's my body type. It's easy for slim people to keep the weight off"*. Instead, make a choice then act on it.

So now choose your lifestyle in the next exercise. Be honest and don't fool yourself. It's not about guilt; it's about picking what you really want. And choose based on what you're willing to start working on today! You don't need to figure out the best running path or find the optimal gym to join. Take small, deliberate, consistent steps like taking the stairs or walking around the block at lunch time. Choose now!

EXERCISE – Diet/Exercise Choice (1 min)

Here are your basic choices when it comes to eating and lifestyle choices. Pick one of the following that is your desired lifestyle (most of the time):

- ☐ Eat anything AND Intense exercise ➡ No weight gain
- ☐ Eat anything AND Modest exercise ➡ Small weight gain
- ☐ Eat anything AND No exercise ➡ Large weight gain
- ☐ Eat smart AND Intense exercise ➡ Optimal body mass
- ☐ Eat smart AND Modest exercise ➡ No weight gain
- ☐ Eat smart AND No exercise ➡ Small weight gain

How much is 100 calories? Two servings of guacamole (4 tablespoons) contain 100 calories. That same energy is also contained in a medium apple (3" diameter). To burn off 100 calories requires walking for approximately 24 minutes. In case you're wondering, less than half of a regular Snicker's bar also contains 100 calories.

The total calories one should consume daily depends on weight, metabolism, age and activity level. The FDA uses 2000 calories as the basis for the Recommended Daily Value (DV) percentage calculation (e.g. saturated fat 6g, 30% DV).

Nutrition Facts Label

Every processed food product by FDA requirement has a Nutrition Facts Label that specifies the contents: total calories, fat, protein, carbohydrates and other information. When reading these labels, be sure to check the serving size. Sometimes this number is lowered just to make the caloric content look better. For example, an 11-ounce large bag or Doritos shows only 140 calories and 8g of fat per serving (12% of the Recommended Daily Allowance). But there are 11 servings in the bag! Seriously, who can eat only one ounce of Doritos. So if you ate the whole bag (in the past I have done this so it is most definitely possible), that would be a whopping 1540 calories or 132% of your daily fat allowance. And Doritos do not taste good with a glass of water so you will be adding more calories with a soda. In fact, one large bag of Potato Chips, Cheetos, Doritos or Corn Chips and 4 colas is approximately 2,000 calories – that is your food allotment for the day!

Burning the Calories

When I do a cardio workout, my average energy burn is 10 calories/minute or 300 calories per half hour. And that's sweating like a pig! If you are 150 pounds and walk on level grade at a modest pace of 3 miles per hour, you will consume 250 calories in an hour[72].

Calorie Counting – Not!

You can count calories and ratio of protein/carbohydrates/fat for every meal – but there's no joy there. Instead, I would recommend you use this knowledge to help guide you to better food decisions. For example, before having the usual soda and bag of chips at 3PM today, have a look at the Nutritional Facts Label to see the total calories. The picture below shows a reasonable 160 calories for potato chip serving. However, always check the "Servings Per Container". In this case the bag contains 10 servings that's 1600 calories if you eat them all!

Nutritional Facts for Large Bag of Potato Chips

Ingredients: Potatoes, Vegetable Oil (Sunflower, Corn and/or Canola Oil), and Salt.

If you had eaten half of that bag and two regular sodas that would be a total of approximately 1080 calories.

Half Bag of Chips = $\dfrac{160 \text{ calories}}{\text{serving}}$ x 5 servings = 800 calories

Two Sodas = $\dfrac{140 \text{ calories}}{\text{serving}}$ x 2 servings = 280 calories

Total Snack = 800 + 280 = 1080 calories

Such a snack (or the calorie equivalent) every day for a month would add 9 pounds to your waistline[73]. Or, if you walk at a modest pace of 3 miles per hour and no snacks along the way, you could burn it off by walking from Los Angeles to San Francisco[74].

1080 calories x 30 days = $\dfrac{32{,}400 \text{ calories}}{\text{month}}$

32,400 calories x $\dfrac{\text{pound}}{3500 \text{ calories}}$ = 9.2 pounds

32,400 calories x $\dfrac{3 \text{ miles}}{\text{hour}}$ x $\dfrac{\text{hour}}{250 \text{ calories}}$ = 388 miles

The number used above, 3500 calories per pound is overly simplified as the calculation does not factor body composition (the percentages of fat, bone, water and muscle in human bodies) so results will most definitely vary by individual. However, the intent of the above math is intended to show that a small change over an extended period of time will result in big changes over time.

It's not hard to know what foods and portions are good for you; just pay attention and make choices. When I've exceeded my ideal amount of food for a meal, I feel bloated, uncomfortable and bad about myself. I now know that these are the consequences when I choose that big fat treat – and I now rarely make that choice.

I have certainly tried diets in the past but without much success. Making a lifestyle change worked for a few years before the weight came back. However, really understanding what was going on in my head and how I used food almost like a sedating drug has made all the difference for me. I have broken through my auto programming. There is no will power to falter, no cravings to keep at bay. I just choose that feeling good about how I look is more important. And there is no pressure because it's not about eating perfectly; if I spoil an appetite and decide to have ice cream, I can get back on track the very next meal – it's one bite at a time!

I don't write down my weight every morning because the tightness or looseness of my clothes gives me a pretty good indication of how I am doing. You definitely know you're doing well if you need to buy smaller size pants. (I had to do this – it's an exhilarating experience!)

Costa Rica

Some years ago I took a 10-day trip to the rainforest of Costa Rica to view the endangered Great Green macaws. It was a wonderful adventure (except for the kidney stone experience) with many discoveries and one completely unexpected one. I stayed at a remote lodge with a limited culinary offering of chicken or fish, vegetables, plantains and salads loaded with heart of palm. Every single day it was heart of palm for lunch and dinner. They harvested it from the property so it was a free source of food rich in calories. To this day I don't eat heart of palm. I often felt the slightly hungry feeling, especially an hour before meals were served. (I would have paid a hefty sum for a bag of Doritos!)

The lodge was situated in the middle of the rainforest with many trails networked out to various habitats where we could watch parrots, toucans, hummingbirds, caiman (South American crocodiles), spider monkeys, sloths and kangaroo rats. We hiked two or three times a day typically 7 miles per outing. It was difficult at first but the scenery was so beautiful and teeming with exotic creatures that before long we all fell into the patterns of our new environment.

When I returned home, I was shocked to see that I had lost 10 pounds in only three weeks! The combination of modest food portions, no deserts and lots of exercise did the trick – the equation works. I remember this somewhat hungry feeling now and know when I'm in this zone. I'm not there as much as I would like to be but I'm now aware of it and make it a choice.

Which one is Better

Now it's time for a quiz. Pick which of the following options you think is the better nutritional choice. (Not the one you would prefer).

- ☐ Cheeseburger with onion rings and a soda OR
- ☐ Chicken salad with sparkling water

- ☐ Linguine with clams and Alfredo sauce OR
- ☐ Salmon with green beans and wild rice

- ☐ Hot fudge sundae OR
- ☐ Fruit salad

You know which one is better for you, right? But we often pick the one that our bodies crave. There's a reason for that and understanding why may help control those bad urges.

Fat is a rare find in the wild. Most of it is found in prey animals and they are constantly running around foraging for food or evading predators so they are ultra-lean. When you happen across a stash of fat the automatic body response is to gorge before someone else can beat you to it. (Reminds me of childhood family Christmas meals when the sausages were passed around). You probably have felt this way when a batch of donuts appeared in the office lunch room. Don't go with the robot response of stuffing down food because it's there – make it a conscious choice.

The structure and internal organs of our bodies have not substantially changed in 100,000 years when we roamed the wilderness in search of calories. During most of the daylight hours, we would look for berries, trap small game and chase (or run away from) big creatures. Today, we can easily get all those calories and more in about 30 minutes by locating the nearest fast food dispensary on our smart phone. Is it any surprise that obesity has become an epidemic in this country? We consume way too much food for the energy we expend. The equation again works – calories consumed minus calories burnt in exercise result in your weight gain or loss. You get to choose.

I try to keep a daily running total of what I've ingested. For example, salad for lunch and no potatoes at dinner with extra veggies means I can have a small dessert!

There are actually three major choices you make at every meal: portion, processing and proportion. The first we have already discussed. The second is the degree of food processing that you consume. Processed food comes in cool containers with fancy graphics and clever marketing campaigns. The other stuff just sits there raw in the produce aisle and behind the butcher's counter. Someone pointed out that the unprocessed food that is best for you is around the periphery of the grocery store – and it's true. Avoid the middle aisles of the store where the sodas, pop tarts and crackers are lurking and you will make healthier choices. In addition to containing less desirable carbohydrates, these foods are loaded with preservatives, sugars, salt and other additives that are bodies aren't sure how to handle so in many cases will simply store as fat. Processed foods have only been around for about 60 years. It seems plausible to me that increases in diabetes, hyper-sensitive allergies and other health issues are directly attributed to our dramatic change in diet. Again you have a choice – fast and cheap food or do you want better quality that takes longer to prepare and costs more? How important is nutrition compared to other items in your budget?

There was an interesting experiment conducted by Baba Shiv a professor at Stanford University[75] in which people were asked to remember a 3 to 10-digit number. But there was a twist. After being given their number, they had to go to a nearby room to recite it. In route, they were interrupted by a hostess offering them a snack or either fruit or chocolate cake to thank them for their participation. Their choice, the actual purpose of the experiment, was to determine if there was a correlation between the amount of data in our head and the choices we made. It turns out there is a strong connection between stress and decisions. Those with the small numbers, the ones easier to remember, tended to choose the healthy fruit while the participants with 7 or more numbers most often picked the chocolate cake. The latter group was much more stressed, needed the reward and had a harder time making the better choice. So be aware that under stress you are less likely to make good food decisions.

Macro Nutrients

Food can be categorized may different ways: calories, vitamins, fats, anti-oxidants, etc. Looking at the simplest level, everything we eat can be broken down into one of three macro nutrients: protein, carbohydrates (carbs) and fats. It is helpful to look at how much of each of these you consume at every meal. First of all, a little background may be useful. Carbohydrates are the first source of energy for the body. In particular, carbs in the form of glucose is needed to keep your brain functioning. Fats are needed for longer term energy and stored away until needed. Protein is what builds muscle. Every time you flex a bicep or calf, that action destroys muscle tissue that needs to be restored — eating protein will fix this.

You can easily find out how many calories, carbs and protein are contained in every packaged product you buy at the grocery store on the Nutritional Facts Label we saw earlier with the potato chips. There are example codes below for raw almonds, cottage cheese and a cookie.

Nutritional Facts Labels

Almonds

Nutrition Facts	
Serving Size: 1oz (28g)	
Servings Per Container 16	
Amount Per Serving	
Calories 160 Calories from Fat 120	
	% Daily Value*
Total Fat 14g	22%
Saturated Fat 1g	5%
Trans Fat 0g	
Cholesterol 0mg	0%
Sodium 0mg	0%
Total Carbohydrate 6g	2%
Dietary Fiber 3g	12%
Sugars 1g	
Protein 6g	12%
Vitamin A 0% • Vitamin C 0%	
Calcium 8% • Iron 6%	
*Percent Daily Values are based on a 2,000 calorie diet.	

Cottage Cheese

Nutrition Facts	
Serving Size 4 oz (113g)	
Servings Per Container 4	
Amount Per Serving	
Calories 80 Cal from Fat 10	
	% Daily Value*
Total Fat 1g	2%
Saturated Fat .5g	2%
Trans Fat 0g	
Cholesterol 10mg	3%
Sodium 320mg	13%
Potassium 160mg	5%
Total Carbohydrate 6g	2%
Dietary Fiber 0g	0%
Sugars 4g	
Protein 11g	22%
Vitamin A 4% • Vitamin C 2%	
Calcium 10% • Iron 0%	
*Percent Daily Values are based on a 2,000 calorie diet.	

Cookie

Nutrition Facts	
Serving Size: 1 bar (78g)	
Servings Per Container Varies	
Amount Per Serving	
Calories 320 Calories from Fat 150	
	% Daily Value*
Total Fat 17g	26%
Saturated Fat 7g	35%
Trans Fat 0g	
Cholesterol 15mg	5%
Sodium 220mg	9%
Total Carbohydrate 40g	13%
Dietary Fiber 2g	8%
Sugars 29g	
Protein 2g	5%
Vitamin A 0% • Vitamin C 0%	
Calcium 0% • Iron 6%	
*Percent Daily Values are based on a 2,000 calorie diet.	

A quick look will show you that almonds contain fat, carbs and protein, cottage cheese is primarily protein and cookies are fat with lots of carbs. Presented this way they may all seem similar but the serving sizes are different: 28, 113 and 78 grams. To get a better comparison of these three foods, the macro nutrient content is shown in the table below adjusted for 113 grams, 4 ounces or half a cup each. The "% Daily Value" is the Recommended Daily Allowance by the Food and Drug Administration in the United States which indicates what percentage of a 2000 calorie diet is supplied with one food serving.

Macro Nutrients for 4 Ounce Servings

	Almonds	**Cottage Cheese**	**Cookie**
Calories	640	80	463
Fat (g)	56	1	25
Carbs (g)	24	6	58
Protein (g)	24	11	3

Clearly the cookie is not good for you. Four ounces of cookies is 463 calories, almost a quarter of the recommended value with a whopping 58 grams of carbohydrates. Almonds have even more total calories for the same weight - 640 calories. So although almonds are a natural snack and good source of protein, small portions are highly recommended. Cottage cheese on the other hand is a great source of lean protein. The same amount of cottage cheese has only one eight the calories of almonds and little or no fat.

In general, you should have roughly an equal amount of fat, carbs and protein at every meal. This will provide immediate energy, stored energy and muscle building protein. From my experience, when I eat roughly equal amounts of protein, carbs and fat I feel satisfied. By comparison, an excess of carbs from bread or sugary foods will keep me snacking all day long ramping up the calories.

Get in the habit of reviewing the Nutritional Facts Label when you go shopping to balance protein, carbs and fats. When looking for a nutritional snack, it is difficult to find protein. So start there and you can easily fill in any needed fat and carbs. While you're at it, keep an eye on foods with high sodium. Salt enables body water retention which can raise your blood pressure and lead to serious medical conditions such as heart attacks, strokes and kidney disease.

Natural Options

It is clear that more natural foods are better for you. The tradeoff is convenience. Processed foods are more available and have a much longer shelf life so with the natural choices you need to shop more often and cook the stuff. However, less processing means fewer chemicals in your body. Natural carbs like vegetables are easier to digest than more complex processed carbohydrates like bread. But even with natural options you need to consider proportions.

For the longest time I believed that you could eat as many fruits and vegetables as you wanted because these were good for you. That seems to be true for vegetables but not fruit. And I would eat a lot of fruit – especially ripe peaches in July! But some fruits are loaded with sugar, fructose to be specific. And although fructose is better for you than processed sugar, it still packs on the calories and potentially keeps your macro nutrients unbalanced. For example, an entire large mango (which I could easily devour in a matter of minutes) is easily 200 calories and can contain as much as 45 grams of sugar[76] – more than most candy bars!

Sugar

And now for a few words about sugar. No doubt you already know that sugar is pervasive in our diets, especially in the form of low cost corn syrup: ketchup, sauces, cereal, creamers – it's everywhere! You wouldn't pour corn syrup in your gas tank because you know it would mess up your car's engine. Why would you pour it into your digestive engine? It does even more damage there[77].

While researching this book, one morning I took a look at the ingredients of my daily breakfast bar which has a balance of macro nutrients: Soy Protein, Corn Syrup – crap! I've been eating these bars for years and number two ingredient is corn syrup; time to find a new alternative. Just goes to show that you must be diligent with what you eat and check the ingredients. I started actually cooking breakfast – 3 egg omelet with mushrooms, spinach and a little cheese. It seems that healthy choices beget healthier choices. And I have found it to be powerfully true for exercise. If I go to the gym, I seem to make better food choices. It may be simply because I don't want to immediately waste the exercise effort. Or it may be that there's something physiological, that your body is trying to tell you what it needs and I'm now more apt to listen to it.

Variety

It's good to mix it up and choose from a wide variety of good foods so your body has a better chance of getting all the nutrients it needs without having to take supplements or making elaborate grocery lists. I have come to believe that fresh is best. Oh and I just purchased a food blender on a friend's recommendation. Seems like a really cool idea; mash up any combination of leafy greens, fruits, veggies, nuts and seeds to liberate the micro nutrients and ingest it all. Having vegetables this way rather than in a salad eliminates salad dressing that can add a whopping amount of calories and fat. Blending a few raw nuts and seeds eliminates all the salt that I don't miss. And it's fast!

EXERCISE – Calorie Cost (60 min)

Would you like to have an 8-ounce regular soda with a 1-ounce bag of chips (300 calories) and not add any weight? Then complete this exercise right now and that will be your reward! Put on your running shoes and go outside for a one hour brisk walk (4 mph), 30 minutes of vigorous cycling, or 60 minutes of fast dancing (aerobic, ballet or ballroom)[78]. Do it! I want you to feel the cost of your choices.

The Sweaty Kind of Exercise

Cardio exercise is really good for you to build your heart[79]. The concept is to maintain 80% of your max heart rate so that under normal conditions, your heart beats slower. Your heart has a limited number of heartbeats so you can get it healthier, it will beat slower, longer and so will you! But the problem is, I find stationary bicycles, Stairmasters and other aerobic machines really boring. For over a year I've been able to get in 20 minutes on the elliptical per session – it's pretty much the minimum to help your heart. I discovered an easy way to stretch it out to 30 minutes – music! If you already know this, great; otherwise, give it a try. Even better, create a song list of your favorite high energy tunes – amazingly motivational. And if I can synchronize my strides with the beat of the music – there's added adrenalin. And it feels like I am one with the universe - try it!

Regular visits to the gym with music can produce impressive results. When I started regular cardio the workout level on the Sports Art Fitness E821, it never exceeded level 5. Today, listening to "Drops of Jupiter"[80] I cooled down to level 12! I was on the treadmill today and the music is so powerful that I stayed on for another five minutes. Yes, the same machine that I used to dread – barely hanging on to the last minute before I could get off. And now, I don't want to get off until the song ends.

Cardio exercise also strengthens the lungs and heart, increases immune system functions, gives you more energy and makes it easier to sleep. So exercise can be a wonderful addition to your new life.

EXERCISE – Health Goal Review (15 min)

Revisit your Life Wheel Goals and review your "Health" goals. I highly recommend that you have at least one. Keeping your body moving is important and a great investment in a long future. If you're struggling, set a low bar, here are some examples:

- 30-minute walk at lunch on Monday, Wednesday and Friday
- One hour playing with the kids in the park Tuesday, Thursday after work and Saturday
- Buy an inexpensive bicycle and ride to the store 3 times a week

Now it's your turn. The key is to commit to a routine so that exercise becomes a natural part of your life. Create a longer life now!

Health Goal 1: _____

_____

Health Goal 2: _____

_____

Health Goal 3: _____

_____

What's a Spiritual awakening?

And how do I get one? Well I will tell you, from my experience, a spiritual awakening is awesome! I define a spiritual awakening as a flash of insight that enables you to understand a fundamental truth that impacts your life in a profound and positive way. When the Grinch eagerly awaits the reaction of Whoville after he steals all their Christmas loot but instead hears joyful singing. He then understands the true meaning of the holiday[81] – that is a spiritual awakening.

Spiritual Awakening Comes from Within

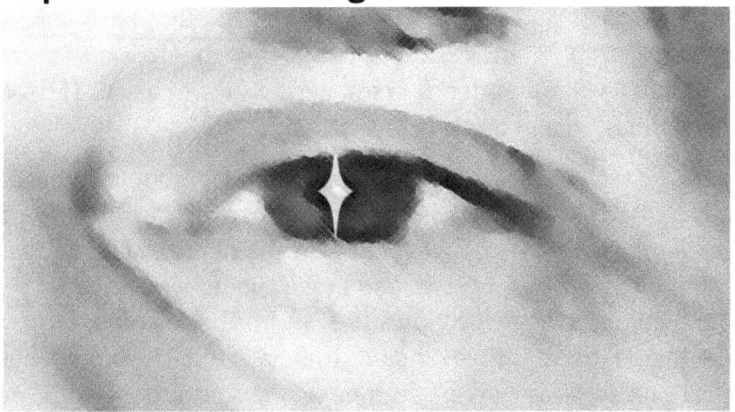

I had a spiritual awakening when I was 52, which in many ways started my adult life. I want to share my experience to possibly help you. Looking at people on planes, in stores and walking along the street many with blank faces completely self-absorbed is how I looked for many, many years. Now I want to run around hugging everyone wishing them "Happy Christmas" like Scrooge in the closing sequence of "A Christmas Carol"[82].

A spiritual awakening is probably not what you think. I read a most surprising story of a revered Buddhist monk who tried to elicit a spiritual awakening with a group of nuns[83]. These nuns had gathered to learn from one of the renowned masters and over the course of three months memorized a complex series of chants. On the fourth month, the monk told them to disregard the previous ones and learn new ones. They were furious! They complained of all the time they had spent to commit the previous works to memory. The lesson however was that it's not about the words, the recipe, or the book. Rather, it was about keeping present, aware and always being open to the world around you – a new way of living. So it is for me as well.

I experienced a series of spiritual awakenings not just one, like little gems scattered in the forest. If I am open to possibilities in the moment, I may stumble across one of these gems and experience a burst of inner joy. Such a discovery usually sets me on a different path in one or more areas of my life.

Living the Moment

My most important life goal is to be a whole-hearted person which to me means that I engage with the world around me, share my thoughts, feelings and truly live the moment – right now. Every moment holds potential to reveal wonder and joy. It is entirely up to us to choose one of them, reach out and participate. It's really not that hard once you realize that such moments exist and you are worthy to engage. Brene Brown coined "whole-hearted" person and has a wonderful TED video that explain the concept and its power[84].

In the past I would spend a lot of time thinking about ... the past; endlessly replaying moments in my head where I wasn't quite good enough, didn't think fast enough, didn't say the right thing, didn't make the best choice and how I wasted so much time. I would justify these negative thoughts as introspection and analysis. Then I would leap to the future with ambitious and clever plans to be a wonderful human being, exceed all expectations and do better next time. Of course, I would need to factor in many scenarios just to be safe – all of which takes even more time. And in the process I would completing miss the moment ... and the next one ... and the next one. The only thing that matters is THIS moment.

The past can't be changed and the future is unknown. It's so simple. Stop thinking and just ... be. Everything else just doesn't matter. I believe that if you fully engage right now, honestly and openly with everyone – the rest will take care of itself. It doesn't matter that you screwed up yesterday, you can be phenomenal right now. And don't worry about the future ... it will be here in a moment!

Shark Story

A good friend shared an insightful story this morning that illustrates this point brilliantly. He was at the beach enjoying the beautiful day, appreciating the warmth, the luxury of just sitting – living the moment. Then he noticed that not far from shore where a rock was sporadically exposed with the undulating swell, an unusual moving organic shape was perched just out of reach of the waves – perhaps a baby dolphin. He looked around to see if anyone had noticed this amazing sight. Although there were a number of people on the beach within easy sight of the event, nobody was paying attention; they were preoccupied with their thoughts and completely oblivious to this wondrous moment. My friend jumped in the water with the intent to rescue the dolphin from its perch. As he approached, he saw that the creature was in fact a juvenile shark who had misjudged the tide. A rogue wave appeared and released the shark from its precarious position who then continued on unharmed chasing the reef fish. Because he was living the moment, my friend experience a rare nature event that enriched his day. That's what I'm talking about!

The Key to the Toolbox

I've been noticing the reaction of my friends when I unfailingly gush with optimism every time they ask me about my day; "Great, Fantastic, Stupendous, Amazing, Astounding, Incredible, Wonderful" are some of my typical replies. At first the looks were a little patronizing; *"Ah, well that will wear off soon"* they seem to say with their knowing nods. Then their expressions grew to disbelief – I had to be making it up. But my sincerity day after day shifted their reactions to a little envy or dismay; *"Hey, I want that too!"* their faces seem to say. Some have even started asking questions: *"How do you find the time for all these fun activities?"*

When I find myself relapsing into old numbing habits, I quickly choose to be active. I hope they are contemplating a change of their own. It is of course up to them, their decision. I can only offer the tools that have worked for me. Actually, I'm providing the key that opens the toolbox; you need to figure out which tools work for you and how to use each one of them for yourself. What I find amazing about this change is that people seem to be actually drawn to me – Wow! So again, I have found by focusing on fixing me, other things like respect and friendship from others, are unexpected benefits.

Enjoy this moment – right now! Don't think, consider, ponder or worry about anything else. A Zen skill that I really like is called: "Mind like water"[85]. As the mental image suggests, in this state you are free from distractions and stress; much like water flowing down a gentle stream unencumbered by rocks along the way, your state of mind is not affected by people or things around you. It is also exemplified by the phrase "like water off a duck's back" – harsh words directed at you from others leaves you completely unperturbed because you simply choose to ignore them.

In my experience, I have discovered that life is not tough, unfair or inherently about suffering as I've heard others bemoan. We impose these bonds on ourselves. So if you find this is your predicament and don't like it - stop doing it already!

I feel a greater strength in myself now that I can articulate my thoughts and desires – an adult confidence unlike anything before in my life, without losing the child's perspective of wonder and fun. It's a new experience – and I like it a lot!

Gecko Animal Guide

My first spiritual awakening, my first glimpse of a promising future, happened when I was visiting southern California on a business trip. I had been looking forward to returning to the state for many years and decided to treat it as a vacation so on the day of departure, I decided to wear my silver Gecko pendant that I obtained on a trip to Hawaii many years ago. Geckos are welcome residents of most homes on the Hawaiian Islands because they seek out and eat any bugs they come across including cockroaches that are their size. Otherwise they are shy and scurry away when you spot them.

After successfully training a roomful of technical sales reps, my host suggested we forego the usual dinner and drinks ritual and instead go surfing; he had an extra board and wetsuit. Although uncharacteristic for me to engage in sports, I agreed. It just struck me as a cool thing to do. After all this was a mini vacation of sorts.

We arrived at a beautiful beach on a lovely spring early evening. After readjusting my wetsuit (I had put it on backwards), I followed my host into the Pacific Ocean. I paddled out … got pushed back … paddled out … fell over … paddled out … slammed by a wall of water … paddled out some more. After about an hour later and nearing exhaustion, I tried to catch my first wave. Clumsily turning the surfboard to face the beach, I was ready for my first wave. My host shouted out to me: "Now! Paddle hard!" So I madly dipped both arms into the water rapidly propelling myself as quickly as I could. And suddenly – wooosh! The wave lifted the board six feet above the trough. I clung to the board with toes and fingers as this force of nature propelled me like a torpedo towards the beach! At that moment I felt such a rush of excitement and positive potential that I knew right then and there that no matter what – I had to move to Southern California.

Inspiration is all around you

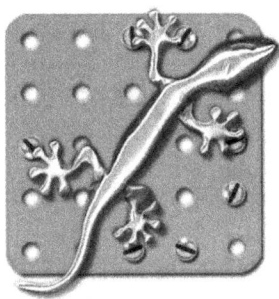

To remind me of this glimpse of a wonderful life, I decided to wear my gecko around my neck every day. Like the geckos in Hawaii, every time I look in the mirror I see that little guy who reminds me to take charge of my life and eliminate anything that stands in the way of my new life. He takes care of the cockroaches in my life by swallowing them whole!

Is there something tangible that has that impact for you? Instead of jewelry perhaps is a piece of art, a pair of shoes, or a photograph. Whatever it is, get your hands on it and put it in a prominent place to remind you of what beckons to your bliss every day.

Back to the Beach

I now go surfing twice as my gecko will attest. I derive such strength, serenity and joy from the beach and surf. It's a good opportunity to disconnect (I do not bring my phone), meditate and get some exercise.

One evening while sitting on my surfboard with the pacific sunset reflecting on the glassy surface of the water I had just felt that warmth wash over me as I recited my affirmation. I had been thinking about a girl who could be a special person in my life. I wanted to ask her on a date but was having considerable trepidation and a fear of rejection. So with the support of the Universe I decided to examine her possible reactions:

1. Yes! What took you so long?!
2. Oh, I'm sorry. I have no interest in you whatsoever. What were you thinking??
3. Hmmm …

The first response is of course my desired answer. The second, although not what I want to hear would provide closure and I could let go of pursuing a relationship knowing that I reached out and there was simply no reciprocated chemistry. The third possibility is that she is uncertain how to respond either because she had not considered my interest or her own in a relationship.

And I found that the third option would be the most interesting of all because I could help her! I could engage her and find out what reservations she has and perhaps alleviate them. Bam! I was suddenly becoming fascinated with the process of connecting with rather than fretting over the fear that I may be rejected or abandoned. It was another little Spiritual Awakening! And my new awareness makes me more confident and comfortable. There is no longer any fear.

Wake up to the life around you

For so long it was all about me and I couldn't see the world around me let alone the people in it. As a child my preference was to hide away in some solitary activity: build electronic projects, take something apart, read science fiction – heck studying was more appealing and safer that engaging with people. No big surprise I ended becoming an engineer. As an adult I learnt some people skills but whenever I had a choice, it was back to my loner activities. As I will explain later, following our childhood patterns is highly predictable.

One day after my spiritual awakening, I was listening to an ancient song by the original crooner, Bing Crosby. The song was "Dream a little dream of me"[86]. After a few lovely verses there is a mellow jazz trumpet improvisational solo by Bob, which still sounds cool 80 years later. In an apparent candid moment, Bing interjects after a few bars: *"Bob! You ain't just dreamin' – you's awake!"*[87]. I could picture him sitting on a tall stool in front of a massive microphone, plying his craft focusing intensely to his intonation and phrasing (he was a perfectionist) and out of nowhere – he notices, appreciates and comments on this special moment he shares with Bob – and us.

Today I was playing a new piece of sheet music on the piano, *"One Flight Down"*[88] sung by Nora Jones. I love it because it is such a vibrant, gospel inspired melody. I have no idea what the songwriter, Jesse Harris, had in mind when he wrote it. It is beautiful and poetic. My interpretation is a message that the joy of life is a song that has always been there – we just need to let ourselves hear it and live it!

To me this was a moment of clarity that helped underline what I had been cobbling together in my recovery – I woke up to the world! Beauty, joy, serenity, peace – it's all there out around us. And all we need to do is let ourselves be a part of it. If this seems trite and obvious, yeah I understand. But all the words that I have been hearing and repeating during these past two years are finally – FINALLY – making sense. Perhaps it's like the process of osmosis, the slow wicking effect when a paper towel is dipped in a solution of water colors. It slowly draws out the individual colors from the murky brown to reveal a crisp rainbow of the original colors.

Innocence

Music has been a great source of inspiration for me in my journey. When the words and music are both aligned to what I am feeling, and I'm paying attention, the experience can be magical.

I believe a spiritual awakening is feeling comfortable in your own skin; when you no longer need to explain yourself or force yourself onto others. Rather, accepting and celebrating you for exactly who you are and not what you do. It is to stop fighting with your unconscious brain and be fully aware of all its aspects, needs and desires. And if ever I feel my programming creeping back, when I don't know why I act or feel a certain way, I stop – and figure it out. Just deal with the issue and move on without agonizing incessantly. In this mode of living I find that the concept of fighting urges just … fades away. Very cool! It seems paradoxical – the more I let go of things and old ways, the more I feel content with my life. You should definitely try it!

If you have seen The Matrix[89], you will recall that Neo wakes up to the world as it really is from the machine induced coma. In order to make the transition he must choose – red or blue pill. I wonder how many people are really awake in the world out there. It feels like I've joined their ranks.

A good sign that you are part of the real world, and out of your own head, is that you engage with it. Before my spiritual awakenings I would often feel self-doubt, self-conscious, self-torment so that it was difficult to engage with people. I now feel like I can walk up to anyone and politely just start a conversion. I'm okay with me and that includes not going to my default tapes if someone says something that used to trigger my auto responses. If you are really good at communicating with people you can skip the next exercise. But if you have any hesitation about talking to a stranger, definitely proceed!

EXERCISE – Meeting People (For Adults Only)

Meet five new people today by just walking up to them and having at least a one-minute conversation. Do more if you are so inclined. These could be complete strangers or people at the office that you may know their names but have to this point only said "Hi" in the hallway. Have a question of two prepared such as asking for directions or restaurant recommendations to break the ice. Jot down your experience below.

    Number of people contacted: _____

    How did you feel before? _____

    How did you feel after? _____

    How did they respond? _____

Do Something Spiritual

How do you feel? Pretty good I suspect. That is the point. Spirituality means different things to different people but we can all agree that it makes us feel better, refreshed. And that's an important piece of a happy life. As stress relief, time to reflect, prepare for the next life challenge or just a mini vacation – it's important.

So now it's your turn to commit to something spiritual in your life if you don't already have a life goal in that category.

EXERCISE – Spiritual Goal Review (15 min)

Revisit your Life Wheel Goals and review your "Spirituality" goals. I highly recommend that you have at least one. A respite from stress and some healing time is what you seek. Some suggestions to consider:

- Walk to the park on every Sunday afternoon without my phone.
- Meditate 15 minutes every morning before getting ready for work with soft music.
- Spend 30 minutes in spa 3 times a week after work before looking at bills.

Spirituality Goal 1: _____

Spirituality Goal 2: _____

Human Connection

Our Purpose on the Planet

Connecting with people – that's it! According to Brene Brown[90] renowned researcher, storyteller and author, that is why we are all here. When we make an honest connection with another human being (does not need to be strictly verbal) we are happy, joyful, whole. When we fail to connect with others, we are miserable.

The next time you see a night cityscape ponder the multitudes that live there or are passing through on to another of countless cities. Each with a rich story not unlike your own that you will never know. Each with desires and fears that they too will tackle in their own way, crafting a unique life. This feeling is beautifully captured in *"The Dictionary of Obscure Sorrows"* under the entry *Sonder*[91].

Everyone has a story

When I first heard of connection I was sure it did not apply to me. But with the discovery of how to be a genuine human being, I began to understand, then believe and finally practice making connections with people. It's amazing when you do get outside your own head how easy it is to just talk to another human being. And most of the time the effort is welcomed.

Right now I'm here at Starbucks and a young lady sat next to me, opened a large book and proceeded to text on her phone. After a few moments I pleasantly commented that she seemed to be avoiding her homework.

She laughed, acknowledged it and we shared a brief fun moment. She did indeed dive in shortly thereafter bravely tackling a challenging biology course 15 years after last being in school - looked pretty tough to me. It really can be that easy to connect with people. And I felt good for having engaged with Maren.

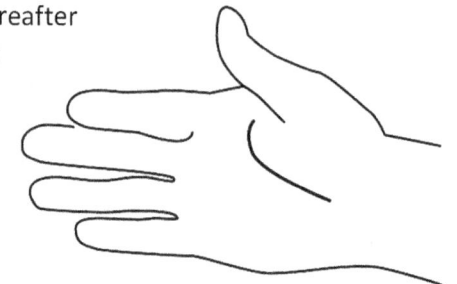

If instead I was sipping my coffee ticked off because somebody took your parking spot an hour ago, thinking that the person next to me doesn't want to talk or worrying about how I'm going to pay the bills next month – I would have missed the opportunity. And it's not only feeling good for having an interaction with another person, there could be really nice surprises. For example, I called Brie (waitress in Colorado from the beginning of the book) to tell her that she will be in the book and she recommended a great restaurant in Baltimore where I was visiting. Turns out she used to live there!

Why do people cave-in?

"Cave-in" is a term a friend of my used to describe how people behaved in a Los Angeles suburban neighborhood after coming home from work. He on the contrary, rejoiced in life, especially if people came to visit so he could have an excuse to show them his fabulous wine cellar and share some exquisite vintages with them. He would regale them with grandiose tales whose truth many questioned but no one really cared – he was so fun! I used to "cave-in" all the time – escape to my fortress away from the world. Sad thing was that I also sealed myself off from my ex within that fortress – I guess you might say the inner sanctum of isolation; not opening up and talking about what I really felt, what I really wanted. Why? Well, my internal programming told me it's not safe to share your thoughts and aspirations because others could squash them and then you would feel horrible. That programming stems from my early childhood and it would not have been safe then to reveal those things in my family.

As an adult, with considerable experience, means and confidence, why did I still behave this way? Because I was still operating with the same programming that enabled my survival as a child. That programming does not change all by itself – it takes recognition, help and a great deal of work. Fortunately, I found that help and wanted joy badly enough to engineer my life into something truly wonderful.

Getting Unreasonable Happiness

I recently read a phenomenal book by Dan Millman entitled "Way of the Peaceful Warrior"[92] (also a 2006 movie starring Nick Nolte) that claims we are all entitled to "unreasonable happiness" and I whole-heartily agree! Do you feel a little embarrassed about wanting unreasonable happiness? Do you want only your share, what is acceptable in polite company? I felt that way. And it's so wrong!! It is quite natural to be happy; toddlers know no bounds and every day they expect unreasonable happiness. We unlearn this as we get older and more responsible - we just need to get back to that early state of being.

First step is to believe that we are worthy of getting happiness. Step two, just get out of your way. What I mean is stop all the roadblocks and excuses we put up to deny the happiness in our lives. It's so easy to blame others for our misfortune or put off happiness with platitudes like: *"sure, I'll do that fun stuff after I get through the to do list"*, *"yes that would make me happy but I have to take care of* <insert person, thing or event here>". You are simply entitled to happiness – period. Accept it. If you're having difficulty with the concept, it's probably because you have been told too many times by others that you're not. You can start to counter that influence by limiting your contact with these negative people, saying a daily affirmation like *"I deserve to be happy"* or even taping such a message on your bathroom mirror. (I've done this, and it really works).

Now I'm frankly a little embarrassed when people ask me how I am and I honestly tell them: superb, wonderful, ecstatic, amazing – and I really mean it. Because it seems they so rarely truly feel that way. Recently, I've had a big setback. But instead of resorting to a night of heavy drinking or scooping to the bottom of a quart of ice cream, I chose to simply acknowledge and experience the feeling – not repressing or fighting it. And, instead of hiding I focus outward on helping someone else. I felt better rather quickly. I have found that when you stop thinking about yourself and reach out to others, in any way at all (opening a door, giving advice, doing an errand, writing a blog) – it really is a great way to help yourself. Try it, you will be amazed - it is the road to unreasonable happiness!

Closed Book

Do you know someone who is a closed book? Someone who guards what they do, what they think, how they feel? I used to be a closed book but no longer because it is completely contrary to living a whole-hearted life, which is my personal number one goal. I want a relationship that is true and honest to the core. In my marriage, we live separate lives. For the most part we were polite, supportive – yet distant. We rarely discussed the important stuff – feelings, aspirations, fears. And yes I was the first to pull up the drawbridge gate when even a hint of such a conversation was in the air. Looking back, it seemed that we pinned our hope on milestones: anniversaries, birthdays and holidays. These were opportunities to do a relationship dump, regroup then take a deep breath before looking forward to the next one and returning back to our respective shells. It was a coping strategy – a way to survive.

A key part of my spiritual awakening was figuring out that joy comes from within. By sharing of yourself, you let me world in but much more importantly, you let yourself out of your own dungeon. I have found that the armor you create to shield yourself from the world around you, and especially negative emotions, does a wonderful job as is a great defensive strategy. The problem is that same armor prevents you from feeling all the good stuff too. I had rarely experienced true joy, happiness or serenity because I was always on alert and ready to do battle - nobody is getting in! I have discovered in the last two years – that is a lonely way to exist. We all deserve a whole lot more than mere survival.

What I want – must have – is a connection to the core with one person so that these milestones become superfluous. Every day, every hour, every moment will be transparent, special and connected. Okay this will not always happen but that is absolutely the goal. Maybe this is where the idea of not going to bed angry comes from? It is the stuff of romantic movies, dreams and ... *I want it!*

Here's my mantra that keeps me focused on making progress with my people skills:

Confidence
Communication
Connection

Confidence

It all starts with confidence. Confidence that I am worthy to have friendships. Confidence that I have interesting things to share. If you have a lifetime of recordings in your head that say otherwise, it is not easy to change your thoughts which is a prerequisite to action. Surrounding myself with people that are supportive and with knowledge of the 90 second rule, I slowly began to erase the tapes.

When I had a bad personal interaction, I always assumed it was something I did wrong. Well, sometimes it's the other person. Confidence will help you sort this out. And when the other person is wrong, be kind because you're trying to build a connection.

I've struggled with understanding what confidence is and how to get more of it. I believe the truth is quite simply – *you are a worthwhile person*[93].

Communication

Once you have a handle on confidence, you can then communicate with the outside world. I now make frequent visits outside my own head – and it is empowering to do so. The process gathers momentum because when you talk to people there is suddenly other thoughts and ideas that are not yours which tends to keep you considering other perspectives and not get locked in to your own loops. Communication is powerful, complex and subtle. For one thing, you may think words are the most important form of communication. Actual in typical daily human interactions only 7% gets internalized from words. Vocal elements make up 38% of communication (e.g. intonation, rhythm) and 55% from body language (facial expressions, posture, hand gestures)[94].

The amount of communication that is nonverbal probably varies between 60% and 90% depending on people involved and the situation. (Giving directions or a stranger or convincing your friend to loan you money will have vastly different verbal vs. non-verbal communication). I love to provide elaborate explanations and was always confused why people's eyes would glaze over after only a few minutes of my discourse; they were reading my non-verbal communication and finding it uninteresting. Now I can read their non-verbal signs and know when to get to the bottom line quickly.

We humans have developed an amazing ability to read people's body language in seconds because it was essential. Language has only been around a short time in our long time on the planet. Our survival depended to identifying friend or foe quickly before we could ask: *"What's the password?"* It is commonly believed that the eyes are the window into one's soul. But this doesn't really make sense because other than pupil dilation under extreme conditions, our eyeballs don't actually change. What gives us information when peering into a lover's or an enemy's eyes are all the muscles surrounding the eyes. There are 43 muscles in our face with 14 immediately surrounding the eyes[95]. It is the particular relaxing, tensing and twitching that we can detect and decode into know feelings and intensions. These responses are automatically generated by the brain and cannot easily be manipulated.

Windows to the soul

In my experience, during first meetings (casual, business or romantic) the words you use do not matter. It's all about making the other person feel okay to be in your company and secondly finding something that the two of you have in common. If all goes well, a second meeting will get a relationship started and the words will play a more important role but non-verbal cues will still dominate the communication.

EXERCISE – Communication Recall (20 min)

Think of a recent memorable communication you have experienced with someone special.

Who was it with? _____

How do you think they felt? _____

What was discussed? _____

What words do you remember? _____

How did you feel? _____

Why was it memorable? _____

Connection

I have learned that once confidence and communication hurdles have been surmounted, I can focus on connecting with the one person before me in THIS moment. I try to listen to them without distraction – really listen to their words, tone and body language; I do not think about my response, nor judge what they have to say but rather, fully experience what they have to say. I try and understand their point of view and be sympathetic but I don't have to agree with them. This is in fact the

Golden Rule: *"Do unto others as you would have them do unto you"* [96] with a modification: *Be the listener you would like to have.*

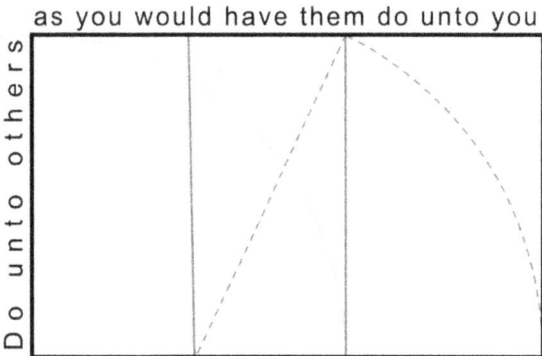

Now this is a challenging thing to do – being in the moment and having a genuine connection with another person. But it is so rewarding! There's no pretense, hidden agenda or competition for who has the best vacation anecdote. To be yourself and have others appreciate you for exactly who you are – truly golden moments. Accumulate a few of these connections with a person you like ... and you start a friendship! Why didn't they teach this in high school rather than history?

Making Friends

For many of you, making and keeping good friends may be natural so you may just skip this section. For me, this was not obvious at all. The whole idea of friends was frankly baffling. At times I felt that I didn't need any – I was completely self-reliant. Other times I longed for someone to share my day. And I was resentful that no one wanted to be my friend.

Good friends are relationships where both parties benefit from time spent together. Choose your friends wisely. People that make you feel uncomfortable, helpless or unappreciated are not friends – move on. There is nothing for you to prove to them. It is they who have let you down.

The most difficult issue after my marriage ended was dealing with feeling completely alone. I had all my human connection collateral with one person – for 28 years. So when that was gone, it was truly a big loss. I felt like the only one on the planet in that situation. So if you feel that way, you're not alone either. And we can do something about it. Start investing time and energy in others by seeking real connections. I will not make the same mistake by having only one person in my life; an exclusive romantic relationship certainly but also a handful of good friends.

You may find this self-evident but for me it was not and figuring it out has been so wonderful, so empowering. Because once again, I can choose. Choose who to include in my inner circle and decide how much energy I want to spend with each one of them. Of course, they in turn have the same option and I respect that without expectation. When we are both in synch – it is indeed a special and healthy relationship.

Last Friday I had "A Wonderful Life" moment. I'm referring to the 1946 Christmas classic movie of the same name starring Jimmy Stewart and directed by the great Frank Capra[97]. If you're not familiar with it, it is a story about George Bailey, a wonderful guy who has many good friends but doesn't know it. On the brink of financial ruin, George believes his life has been completely meaningless and heads to the local bridge to end it all. His imminent suicide in intercepted by an angel, Clarence Odbody, who shows him how his community would have turned out if George had not lived and helped so many people – that were indeed his good friends. Even though in black and white, highly recommended viewing – it ends well.

Last Friday I went to a farewell party for a former colleague who had been recently laid off. After retelling some stories and sharing observations about the company, he shocked me by sharing how much I had impacted his life in the year that I had known him. He was especially grateful for a conversation we had at the company Christmas party where we truly connected. I was floored. Me? I made a difference in his life? I have been focusing on connecting with people to fix me and feel like a real human being by engaging with the world. Sure, I've been helping others when I could. But to actually have an impact on their lives – that is an amazing feeling! And I can truly say that I did indeed FEEL the emotion – very cool!

I am most fortunate to have a wonderful friend with whom I have the privilege of speaking with regularly. We can and have talked about everything, especially the core stuff that matters. He has inspired many of the topics in this book. I attribute this wonderful friendship to two things: we listen supportively without judgment and we enjoy the time together without expectation. It seems the stuff that works is simple.

Making and Keeping Friends

Don't try to be everybody's friend. If you start by focusing with a connection, you can only establish real friendships with a small number of people. Do you really need 100 friends? (By the way I'm talking in-the-flesh friends not online-profile-people that some rack up like frequent flyer miles). Friends require time and that is a limited commodity. Be sure to plan adequate time for your friends to keep those connections going.

Treat friends and in fact all people, the way you would like to be treated. Again this is the Golden Rule. We should carve this in stone on every street corner because it's such a simple idea and really easy to implement if only we remembered to do it more often. All you have to do is consider how you would react to your own words and deeds if you were the recipient.

Go to places where you are likely to meet like-minded people with similar interests. It seems really silly to spell this out now that I see it in print but again this was not obvious to me. I started taking ballroom dance lessons met many people with an easy conversation starter every single week: *"Would you like to dance?"*

Now for a twist on the Golden Rule; want to meet interesting people? Be interesting TO THEM. Want to have someone to talk to? Lend an ear to someone else. I have found that once you take care of yourself, begin developing some confidence and take an interest in others, people will then be attracted to you. It really seems so obvious from this side of the street. All you really need to do is just out of your own way.

Hang out with people that "get it" – people who look outward and want to have genuine relationships. Avoid the rest; I have learned to quickly recognize these people; they make me feel drained and unmotivated after even a short conversation. At times I was one of them, but now most of the time I take responsibility for keeping the conversation going and try to empathize or lift the other person.

"How to Talk to Anyone" by Leil Lowndes[98] is a crafty and powerful book containing 92 little tricks for big success in relationships. She captured the idea of connection brilliantly in her introduction. To paraphrase, when arriving in a room full of people, don't think of how lucky they are that you have arrived, but rather the exciting people you are about to meet.

I believe what you find alluring in others and what they find attractive about you, revolves around how you feel about yourself. After you get to know someone, they become part of your circle and become a select subset of the sea of people out there. Somebody said you need to meet 100 people to find one that you genuinely connect with and that concurs with my experience. So pick activities with lots of people, especially stuff you like to do – and get out there!

The meaning of life is connecting with another human being. Finding a few people that you can relate to is what it's all about. Sharing brief moments with them makes life joyful. Try it!

EXERCISE – Get to know someone (60 min)

Think of someone you would like to get to know a little better. It can be for business, friendship or potential romance. You can share a lunch with a co-worker, invite a neighbor over for coffee, or ask someone you really admire and respect for advice.

Who do you have in mind? _____

Reason for meeting? _____

_____

Topics you have in common? _____

_____

When will you ask them? _____

Alienating Friends

Once you make friends you need to treat them well. If you're not up front with people on what you want and how you feel, there's a cascade of bad things that can happen. Resentment can quickly build in your head taking up precious time, energy and instantly kill joy. It can put you in a bad mood and the feeling that the whole world is against you. Avoid the drama and bad feelings – just be honest with others on what you want and how you feel – they may surprise you.

Kids

I have no experience here other than being one. I do however have a few thoughts. Children must be kept safe, nurtured and deserve your time and attention. Don't forget to ask for help if you are the primary care giver and be sure to make time for yourself. Remember the airline safety video, if the cabin depressurizes put on your oxygen mask first then attend to the children. Because if you run out of air you will not be any good to your kids.

My therapist told me with dismay of a family event where his nieces were texting each other – while sitting on the same couch! And it seems that this is not a unique situation. When these soon to be adults are transfixed texting on their portable devices, they miss connecting with the people sitting right next to them. Those subtle but vital skills like reading facial expressions to understand when someone is frustrated, sad or angry may be lost without regular face to face contact. I fear the potentially dire consequences in the future of our inability to live peacefully together.

The evidence suggests that this erosion of communication skills is happening with kids across the country and around the world. Not only are these children not developing their human interaction skills that will be crucial when it is time to find a job, they are severely handicapping their neural development. Jill Bolte Taylor has an excellent video that summarizes the drastically new view of how the brain evolves during the teenage years[99].

It is critical that these kids experience as wide a range of activities as possible between the ages of 13 and 19 because they will lose half their brain cells! The research shows that in the massive neural growth spurt in the pre-teen brain is trimmed based on synapses that do not have significant connections. So if you are not exposed or sufficiently experienced in a particular skill or thought process by age 19, those cells will be jettisoned. And learning how to effectively connect with another person without the use of a screen later in life will be far more difficult. (I know, from personal experience).

EXERCISE – People Goal Review (15 min)

Revisit your Life Wheel Goals and review your "Friends & Family" goals. I highly recommend that you have at least two, one for family and one for friends. Both are essential to a fulfilled life because we all seek connection with people. Goals can be either specific to an individual like "take mom shopping at her favorite mall once a month" or more general such as: "be friendly to 5 strangers every day".

People Goal 1: _____

_____

People Goal 2: _____

_____

People Goal 3: _____

_____

People Goal 4: _____

_____

Romance

The Courtroom

I'm in the hallway outside of Daphne Syke Scott's courtroom awaiting the hearing that will finalize my divorce. My ex just arrived, I said hello but she just smiled curtly and sat across the hall from me. It's a surreal moment, one that I have not been especially looking forward to – okay ... dreading actually. This is not how I envisioned a 28-year relationship would end. I guess that's the nature of co-dependency – all or nothing at all. (Also known as black and white thinking in ACA[100] terminology). In hindsight this moment highlights the importance of shared moments; since there has been no substantial contact between us for a long time, there are no recent shared moments and there is only a void to grieve. This lesson of the importance of shared moments I will remember.

I know with certainty that turning over my choices to another person will not ever happen again. To fully enmesh with another so closely that all sense of self is blurred beyond recognition is not a healthy approach. To live "Inside each other's pockets" sounds charming but in reality is a self-imposed prison keeping out the rest of the world. Like the Cypress and the Oak that grow in the same shadow, their roots become so intertwined they both lead stunted lives in each other's space.

Thinking back to the very beginning of meeting my ex, the first date in fact, I had mentioned marriage. Since I had spotted a co-dependent person that was so comfortable, it was a done deal. As a result, my proposal was lackluster – it actually seemed redundant to me. These are insights gained after a great deal of work in recovery. At the time it seemed like wonderfully falling in love that I had heard, seen so many times before but never felt. My ex would reprimand me repeatedly for this poor proposal – and rightly so. I just didn't know what I was getting into nor what I really needed. But next time, whenever that opportunity presents itself, I will enter it with full awareness, enthusiasm and consideration of my partner.

What is Love Anyways

One of the first books I read at the beginning of my journey to figure out what was going on in my life was "Struggle for Intimacy" [101]. In it I found to my utter astonishment that many other were confused about love. I felt a glimmer of hope – it seemed that I was not alone in my wretchedness. I felt shame that I did not understanding love – doesn't every person get it? It turns out I had a great deal of company in my dysfunction. Now I understand that I simply never learned how to love me so I couldn't possibly love someone else. After acknowledging that I am worthy, working hard to change my programming and looking outward, I figured out what love is and how wonderful it is to have it in your life.

A New Start

With a potential new long term relationship on the horizon, I am aware of the dangers of returning to my programming: not feeling worthy enough to ask for what I need, assuming the other person is always right. Instead, we will share the experiences, challenges and joys together. From this point in my journey it seems so obvious.

It's amazing to me how some principles apply to many areas in life. For example, the concept of "letting go" when dealing with frustrations at the office, can also apply to your love life. I will just let it go and instead enjoy the discovery of who she is in the moment without judgment. It is rather easy if you just stop worrying about you and consider the other person.

I believe you need to find a friend first then one may turn into a soul mate. Finding people that you can connect with is critical. But a soul mate is a level more wonderful and much more elusive. I believe that starting with simply connecting with people honesty is a good step to a potential romantic relationship because it forms a solid base on which to build.

This new insight has really helped me see the other person's perspective. When I find someone interesting but they don't reciprocate the interest, it's okay. I now understand the eagerness when someone finds me attractive and I can dissuade them gently with compassion. With this approach there's no room for feeling rejected or unworthy – it was simply not a match. Ah ... so much better.

Blissful Biology

Physical attraction is certainly a factor in this endeavor. Our biology wires us this way and it feels so intoxicating to meet someone who may be "that special person". I had such an experience recently and although it is a modest beginning, it is so cool! I feel alive with joy and possibilities! I heard a Dean Martin tune last night from 1956, "On the street where you live" from "My Fair Lady"[102]. To summarize, the song is simply about a love struck guy who is wandering aimlessly down this special street in town singing to the world that this is the only place he wants to be – on the street where his love interest resides. He doesn't even see her! It's good enough to know that maybe, just maybe she will appear. Actually, it sounds like he spends all day walking up and down the street based on the looks he gets from the neighbors. So the odds are pretty good that he will see her. This is not a strategy that I have used nor is it recommended. But it does convey the feeling so well. While listening, I was singing aloud and bopping like Calvin & Hobbes[103].

I now really understand why there are so many songs about finding love – it is such a fabulous feeling. As you may recall from my colored Life Wheel, finding romance is currently my life goal priority. Yes, things may not turn out with the lovely lady who has mesmerized me with her beauty and intelligence, but I'm out there embracing life! And no matter what happens, I am proud of that big change in me.

Beautiful people may draw coveted, admired and lustful glances but once you start talking with them, the physical characteristics are not so important. It's within that matters - what we connect with – beauty becomes secondary.

Soul Mate

I think keeping the romance alive is a big challenge and essential to a healthy partnership. Otherwise all the little things and perhaps some big ones can start to accumulate and build a wall between the two of you. Someone once told me many years ago, to deal with those small issues as soon as possible or else they will grow. Sadly, that was so true in my case. To prevent this from happening again I have been working intensely on fixing me and developing one on one communication skills, specifically verbal Intimacy. That is sharing with the other person what's going on in your head – how you feel, what you really want, what you fear. Guys, I suspect many of you struggle with this too. I come from a lineage of peasant stock where grunting was considered sufficient communication. So if I can do it, you all can as well. Pay particular attention to the following sections in this book: "Confidence", "Dysfunctional Family of Origin", perhaps even "Addictions".

Friends and family are important to diversity your human contacts, not unlike advice commonly given to diversify your investment portfolio for financial stability. I found that it is not healthy to have only one person in your life who is best friend, lover, confidant, business partner and roommate.

Seven Year Itch

During an episode of "This American Life" with host Ira Glass I heard an intriguing 7-year marriage concept[104]. This is a challenging alternative idea to either being single and distancing yourself from any kind of commitment or married forever.

The idea is to have a marriage contract that is up for renewal every 7 years. Unless both parties agree to re-commit, the default is an automatic end to the marriage. Such an agreement would give both partners the attention they deserve and I believe dramatically reduce unhappy marriages. It certainly would be an inducement for both partners to avoid taking the other for granted and provide an opportunity to review how they have been treating each other.

In my case it would have been a much healthier option. Once married, I just checked it off my list and moved on to other priorities. I did not think that it was important, to keep it central to my life goals and make changes as needed.

Here's a crazy idea! If you have a significant other, both of you write down your top 7 life goals and share them. (Best to write them down independently so that neither of you are influenced). I would expect an animated discussion after you review your partner's goals and hopefully you have a few in common. For the goals that are different, you can talk about what is important and requires time away from each other. This can be a healthy thing, especially if you negotiate honestly and up front. These are opportunities for you to share your vulnerabilities and connect in a deep way with your partner. If there are no surprises when reviewing your lists, congratulations you have an awesome relationship!

Right and Right Away

The prospect of a new long-term romance is a heart-warming thought for me. And this time I will get it right; I will consider the other person and myself every day. It will not be perfect because sometimes I will get it wrong; that's okay as long as I fix it right away. These words are easy to say but much more challenging to truly understand. Looking back over my years of marriage I now really understand how vital it is to let the other person know what is going on honestly. (Of course that assumes that you know what is going on with you. I needed to figure that out first). There is absolutely no benefit to keeping your feelings, thoughts and desires to yourself. The other person does not know what your issues are; you may feel frustrated; the stuff between the two of you will build over time and NOT fade away. If your goals no longer match up with your partner's, in my opinion it is far better to try to settle them or end the relationship.

I met a lovely woman recently that has opened some intriguing questions in my quest for a better life. She is not available but would otherwise be the recipient of my gallant attention. The complexity arises because I found out she is an occasional smoker. In the past I declared that all smokers are followers of Satan at worst and uncaring, undisciplined cretins at best. This stems from bad childhood memories. We lived in a remotely located house for some years with only one bathroom. My father would be first in and fill it with smoke while coughing up a lung which let me know it would soon be my turn. During the winter it was forbidden to have the windows open so I just had to endure the secondhand smoke or burst a kidney. (Peeing in the snow wearing only pajamas was frowned upon). But now that I have mitigated my extreme views, would I consider dating a smoker? And if so, would I try to help her quit? Don't know, but I think it is healthy to ponder these questions.

The larger question in seeking that special someone is: should I consult a checklist or just experience who they are without any requirements? Hmmm ... the former sounds judgmental while the latter seems froth with peril. After some deliberation, my inclination is to look for someone meeting these guidelines:

- Healthy in mind, body and spirit
- Knows who they are and does not NEED someone
- Someone who puts my Soul Smiling Quotient through the roof!

Seems like a good place to start. When I did not express my thoughts and just kept trying to do what I thought my spouse wanted I began to be resentful, withdrawn and unhappy. To stay in that situation, I was inadvertently programming myself to believe that this was as good as it gets so I better just get used to it – which of course came true. Is it hard to share your feelings? Perhaps you're not even sure how you feel about some issues? That was me. I would strongly recommend that you figure out why. The section on "Dysfunctional Family of Origin" may be a good place to start.

You are in charge of your feelings

I have learned that each person is ultimately responsible for their own feelings. In a relationship probably the best thing you can do is just be there to listen to your partner – don't suggest, don't try to change, don't fix them – just let them experience the feeling. You may even gain further empathy for your partner. And in turn, they may return the support in kind. (If they don't, you may want to gently ask for their support in this way). I believe this responsibility of feelings is indeed a fundamental requirement for a meaningful, satisfying and long-term relationship.

I'm now ready to find that special person in my life and can see the possibility of a healthy relationship. Something that for all my life was so baffling, foreign and elusive.

Online Dating

To celebrate my finalized divorce, I signed up on one of the popular online dating sites. My perceptions were that dating sites were stilted, disingenuous and time consuming. To date, my primary means of looking for a special person was at the dance studio – fun exercise and socializing with dozens of women seemed like a winner plan. Sadly, after 6 months and dancing with 300+ women, still have not found her.

Online dating is quite different compared to the dance studio experience where you say "Hello" then clutch each other in a sultry Tango hold. The dating site is from the inside out; you find out what motivates and impassions your potential match long before you even hear their voice, let alone get to dance with them.

So I've setup my dating profile, complete with 12 pictures; what a great exercise! It forced me to figure out what I really want in a life partner – never done that before. My choices in the past were driven by my programming – fear of abandonment and attraction to familiar dysfunction. But now I get to choose – and it is so exciting! Despite the fact that I haven't dated in a really long time, I am looking forward to meeting someone to share common experiences, joys and alchemy – the spark that will ignite a glorious relationship!

The online service that I opted for uses "Guided Communications" to slowly get matches acquainted with each other. Here's how it works; complete a profile (took me about 2 hours), including your criteria (age range, distance, ethnicity, etc.) and then wait for suitable matches to show up in your folder. Bink! I got one! Now you review their profile including what they are seeking in a match.

Reviewing matches is a little daunting - where do you start? Pictures, sure I must admit that is my first stop. Attraction – of course, need to have that. But I've discovered that photos can also reveal emotional health which is really important to me. There are clues in body posture, eye contact, chosen background, smile, sun glasses and other people included in the photos. Then I review profiles for passions and priorities. If you're interested in say ... Linda, you send her Five Quick Questions from a list of 50 and wait for her to answer. Linda now has the option to answer the questions then ask five of her own, banish you to the "Hidden" folder or block you – which means her photo goes "poof!" in your folder and she vanishes from your world. If all goes well and you complete her questions, you move into the "Make or Break" stage, where in an equally slow process you share your top 10 things that you must have in a relationship and choose 10 deal breakers from an extensive list. Now you can each ask three of your own "Digging Deeper" questions. If you make it through those then you have access to email each other via a secure private mail system.

With just a week of experience on the site, the waiting for a response can be painful. Kind of like a Band-Aid, if I'm not what you're looking for – rip it off already! This seems to be a fairly common situation based on my limited research and it does make sense if you think about it. First, newcomers have nothing in the queue whereas those that have been using it for a few months have on-going conversations so they are juggling a few options. People do have lives and are often distracted and unable to respond in a timely manner. And these checks allow people to proceed slowly and consider their choices carefully – a really good idea when you're looking for a life partner. (I'm also hoping my compulsive checking for new matches every couple of hours will abate soon).

There is however an option to speed up the glacial pace of "Guided Communications". You can at any point go directly to the proprietary email. But if the other person prefers the slower approach, they can reject receiving the email.

Four months into the experience, I've gone through the process over a dozen times ending at various points along the way due to incompatibility. There has been a few actual face to face dates but just one or two per person – not quite right. Then on Valentine's Day I had a third date planned with the same woman – things were looking up. Found a picturesque hiking trail, picked out a lovely lunch spot and had

167

chocolate hearts ready to go. It was a lovely day in southern California, we were both on time, conversation was engaging, she laughed at my jokes, the meal was great. Back to my car to share the valentine chocolates and ... nothing. We both realized that we could be the best of friends but there was simply no chemistry.

It was sad and I resolved that the dating site approach was simply not working for me. Would I ever find her? Was she even out there?

Downtown Disney

So it was Valentine's Day with no plans for the evening. Reluctantly, I decided to act on a friend's suggestion to check out a live band swing dance at Downtown Disney. I found the outdoor dance floor and 6-piece ensemble perched atop a traditional bandstand in the center of the venue. The floor was packed and surrounded by admiring park visitors. I just waded in and immediately spotted some friends from the dance studio and starting dancing – it was great! First time dancing with spectators – really cool because I remember being on the outside looking in just a few years ago; and now I had some respectable swing moves. Then my night became magical.

I was looking for my next dance partner, turned to my left and saw ... Serrah. There are a handful of dramatic moments in one's life that elicit uncanny clarity, joy ... and change everything. This was such a moment.

She returned an intrigued head tilt and said I looked familiar. I had no idea. Believe me, I would have remembered having this goddess in my life. Fortunately, it is now second nature for me to ask a woman to dance within earshot of any music with a decent beat and a makeshift flat surface – we were on the dance floor in a flash.

And in just two steps ... it is so clear to me that Serrah is an amazing dancer, delightful, dazzling and oh so sexy! Honestly, the next hour was a blur; I could not believe this amazing woman kept dancing with me. But I do remember how I felt – joy and happiness flooded my whole being. Suddenly, I understood so many love songs at a whole new level. But all too quickly, she was gone.

About an hour later, I realized to my horror that I didn't ask for her number. Have you seen "Streetcar Named Desire"? Well, this is how I felt: *"Seeeeeerrrrrraaaaaaah!!!!!"* I had no way of contacting her. And even worse, I was about to go on a two-week business trip so could not scour all the West Coast Swing venues in Orange County.

Back at home with the little I knew about her, I began investigating ... checked out social media sites and found about 30 potentials that were likely to be her (some profiles don't have pictures) and sent the following:

> "Looking for the dazzling and delightful Sara that I danced with on Valentine's Day at Downtown Disney. If you are not her, I apologize for interrupting your day. If you are that special Sara - we need to dance again! There was an amazing connection on Saturday night ... The magic ingredient for a great partnership.
>
> Steve"

She responded!

EXERCISE – Romance Goal Review (30 min)

Revisit your Life Wheel Goals and review your "Romance" goal. I suspect this will be a challenging one for most of you. Give it a shot anyways! Life with romance is so much better. Your goal can be small, or it can be big but I suggest you always have a romance goal on your list. For example:

- Surprise my significant other with a small surprise gift or activity every Sunday.
- Go to one new place every weekend where I can meet men/women my age.
- Start planning a special birthday, valentine's day, anniversary for him/her NOW!

Romance Goal 1: _____

_____

_____

Romance Goal 2: _____

_____

_____

Romance Goal 3: _____

_____

_____

Home

Recently becoming single has made me aware of aspects of life that I had largely ignored in the past. One of these is having a "home". This was my ex's domain, and I suspect that is true for many other couples as well. Sure, I would work on home projects but I treated them in isolation, checking off the "To Do" List and moving on. I rarely appreciated the effort and effect of designing your home space. But now that it's all up to me, there's no choice but to notice the stark difference.

I made a shocking discovery after looking under the bed for the first time in 6 months – a veritable forest of dust bunnies! I'm tempted to setup a time-lapse camera under the bed equipped with night vision, as I'm pretty sure they didn't get there by themselves.

Feng shui

The concept behind the ancient art of Feng shui (literally wind and water) has origins in Chinese astronomy 3500 years ago, to encourage good energy or Chi in your space while eliminating or minimizing the bad energy[105]. The Bagua on the next page is oriented to the cardinal directions of your house and each of the eight areas (fame, relationships, creativity, etc.) correspond to specific rooms. For example, add a plant to represent wood in your "wealth" room or a silver metal flute to encourage creativity in your study. The elements can help you select appropriate enhancement objects by material and/or color. The "earth" can be represented with brown colored objects or pictures. Use blue or teal where you need some "water" power. The "wood" element calls for browns and greens if trees (e.g. bonsai) or plants are not practical. Use vibrant colors of red, yellow, orange or pink to add "fire" to kick start some more fame or bolster your reputation. And gray or silver can encourage travel in your northwest corner if you can't figure out what metallic object would fit your décor.

Bagua to enhance your home and goals

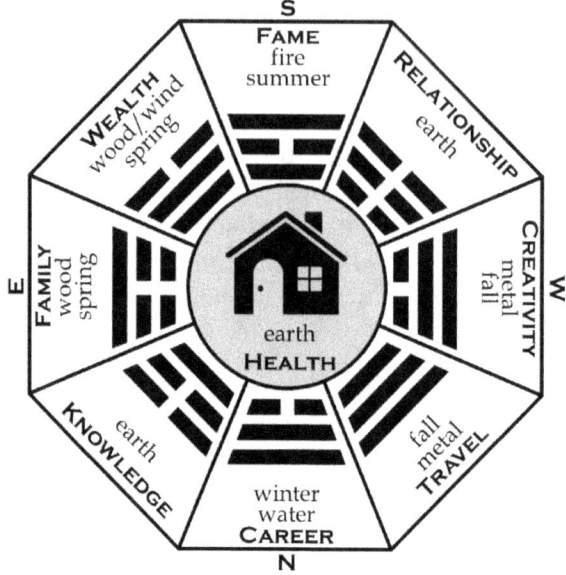

I believe the biggest value of using Feng shui is to load up your house with specific intention cues that are meaningful to you. If you wake up every morning and see your running shoes on a brown doormat that looks like "earth" you are more likely to go running and meet your "health" goal.

Feng shui encourages you to get rid of excess or non-functioning stuff in your space which can have a profound consequence of de-cluttering your life. Fix or get rid of non-working appliances, tools and electronics. Make a decision on projects that have been lingering around the house for a long time – complete it or get rid of it. The constant reminder of the projects you have been neglecting are draining. There's a reason you're not getting around to it. Deal with it then you can focus on new adventures with a clean slate.

I had a massive glass project (over 300 pounds) that was boxed up for a few years and sitting in the garage mocking me; I had a physical reaction every time I saw the crate reminding me of the failure inside. One day when I was in a Feng shui spirit, I cracked it open. A massive glass mosaic of a desert landscape measuring 4 feet wide by 6 feet tall. Constructed with one-inch thick Dalle de verre, the same glass commonly used in churches. Beautiful but was in need of major work. In preparation for a house move, I had rushed packing the artwork under the time pressure of the closing date – and damaged it. It was in a dozen pieces, many completely torn from the polymer matrix, with some completely shattered. Now I was facing my failure head on and focused on making this right even though it seemed like an unsurmountable challenge.

After researching techniques and suppliers, I gained the confidence that I could do tackle the project. Determined that the best strategy was to cut the large piece into quarters so as to reduce the total weight per section yet still preserve the beauty. I obtained glass supplies: slab glass, epoxy, sand and 2-inch steel channel for the frame. I wasn't taking any changes – this would be well engineered. After solving some construction challenges and three months of diligent work, it was done! It turned into a beautiful living room panorama rather than a tormenting eyesore lurking in the garage.

Dalle de Verre Dread to Delight

Don't let things pile up. Deal with it now and be done with it so it's not rattling around in your brain; eliminate draining distractions so you can enjoy your home a little more – even if that's just enjoying the view while having your morning coffee.

If you have items that are broken, unused or just plain don't like, take care of it by fixing, selling, donating or throwing it away. You will unburden yourself and feel much better in your home space. Can't quite get yourself to do so and want to just put it off? Take a moment to figure out why. Has your programming just kicked in? I challenge you to do the next exercise!

EXERCISE – De-Clutter (60 min)

Walk through each room in your house and jot down any items that induce unease, annoyance or make you sigh. Then prioritize the items starting with the one that is most mentally draining for you. Now with the top 5 items, decide right now what you will do about them. And if you can, act on those decisions right now!

PRIORITY	ROOM PROBLEM	FIX
_____	_____	_____
_____	_____	_____
_____	_____	_____
_____	_____	_____
_____	_____	_____
_____	_____	_____
_____	_____	_____
_____	_____	_____
_____	_____	_____
_____	_____	_____

I apologize ahead of time if this exercise inspires some painting or a bathroom remodel. However, in the end you will have a nicer home and that can't help but contribute to a happier life for you.

Finding Your Home

This was one of my life goals that came late in the writing of this book. The delay was primarily due to the uncertainty as I was figuring what was important to me. Buying a home is a big commitment and with experience purchasing seven homes, I have discovered that taking your time is a great approach. After ending my marriage, I rented for three years. At first it was nice not mowing the lawn or fixing broken appliances – gave me some extra time to do other things I enjoyed. However, I missed being the king of the castle and just deciding to do some minor construction or hang some artwork. It was time to find a new home.

I was amazed at how many loan officers and realtors said to me: *"You know ... you can afford more house."* To me this indicated that most people looked for a house at the top of their financial range. Before starting my search, I determined my house requirements. Then I decided what I was comfortable spending monthly on my mortgage which set my maximum house budget. Again it is about balance. For me, it was important to have money for going out to dinner, having occasional lavish weekends and the ability to deal with unexpected expenses (like new car tires). Sure I could have managed an additional $500 to my mortgage to have purchased $100,000 more house. I chose instead not to have the added pressure and a little more on-going luxury. It's your choice. Just make it a deliberate one.

As with most projects, start with defining what you want. Do this before you go on open houses and fall in love with a house. Without understanding your needs, it is difficult to know when you've stepped into the right home for you. If you do the following exercise after you've seen your perfect house, more than likely you will fool yourself: *"oh, those train tracks are not really that close to the bedroom"*. Let's craft your home spec!

EXERCISE – Home Specifications (60 min)

Here are some specifics to consider before you see your first open house:

WHY
- Why are you looking for a new home? _____
- Why are you looking now? _____

WHAT
- What lifestyle do you want? _____
- Buy or rent? _____
- Condo or homestead? _____
- Fixer-upper or ready to move-in ready? _____

WHO
- Who will be living here? _____
- Do you need visitor space? _____

WHERE
- What city/area? _____
- Downtown, suburbia or rural _____
- Traffic to work at rush hour? _____
- How close is shopping? _____
- Close/far enough from family? _____
- Restaurants? _____

WHEN
- Move in? _____
- Event driving your decision? _____

HOW MUCH
- How much space do you need? _____
- _____
- Max Budget? _____
- Desired Price? _____

Now that you are armed with dream house details, go out there and see houses – lots of houses! Review your requirements after seeing your first three houses and determine how they stack up. Revise your Home Spec if needed. Consult with the family to solicit inputs, especially from all those that will be living there. Do this repeatedly and preferably without a rigid timeline. From my experiences, I made the best decisions when there was no pressure of selling my old house before buying a new one.

I found it helpful to rent for a year or two before buying, especially when moving to a completely new area. It was easy to leisurely get to know the different neighborhoods, especially near my workplace and favorite spots to strategically identify the best area for my new home. And when it was a good time to buy, the lack of a selling contingency made me a much more attractive buyer. In two cases it gave me the edge in multi-offer situations.

EXERCISE – Home Goal Review (30 min)

Revisit your Life Wheel Goals and review your "Home" goals. We all enjoy a nicer home so this one should be pretty easy. Here's a few suggestions:

- Repaint the house! Get samples and invite the whole family to a paint party; everyone suggests colors for each room and paint sample swatches on the walls. (If they do a good job, invite them back to do the whole room).
- Upgrade master bath with Jacuzzi and double vanity.
- Check prices for 2 bedroom homes closer to ork and get estimates of value of your house to see if moving is an option.

Home Goal 1: _____

_____

Home Goal 2: _____

_____

Home Goal 3: _____

_____

Money, Job, Career

Simple Concepts that helped
You earned it. Make sure you spend it consciously. If you haven't yet earned it, don't spend it.

Don't be manipulated by high pressure salespeople. When in doubt, delay your purchase and give yourself time to think about it. Do you really need it? Can you afford it? What are you trading for it? Does it support one or more of your life goals? Salespeople understand what's going on and will do whatever they can to keep you off-balance, steer you back to your programming. (Why do you think they're called "hot" buttons?) They know if you think about it, you probably won't buy it.

Another approach is to go to the store prepared (mall or website). Research the product, know the facts, learn competitive pricing and your budget. When I go grocery shopping, I try to keep an approximate running total in my head of what's in my cart. I use this to make better choices when I reach for that $6 box of cookies: *"hmmm ... maybe this is not a good money or calorie choice"*. I'm also alert at the cash register to how much the total should be so I'm not surprised. If it is more than expected, perhaps there's an error which I can address with the cashier immediately. And it's a good feeling to be living consciously and in charge of my day.

I have discovered that the secret to negotiations, when bartering in Vera Cruz or buying a car, is to want it less than they want to sell it. That means you may not get it today but you may get it tomorrow at your price. It usually comes down to time or money. So make it your choice and decide what you want.

Financial Health Equation
Much like the equation that controls your weight, there is a simple equation that controls your financial health.

$$\text{Income} - \text{Costs} = \text{Financial Health}$$

You want "Financial Health" to be greater than zero. That is to say be "in the black" rather than "in the red". The latter idiom refers to an old standard accounting practice of using red ink to indicate a negative balance. I remember as a kid being dragged along to the local bank, looking up at the intimidating bars separating my mother from the clerk as she slipped her bank book under the sill. The teller would use a massive typewriter to hammer in the bank balance whose keys seemed to echo in the stark room. Stumbling through the revolving door on the way out, she would peer earnestly at the addition in the checkbook and curse if it was red. You definitely want to be "in the black" - income greater than costs; which means you are actually building wealth or saving!

In seeking happiness, you may find your ideal job pays a whole lot less that what you're making now. Instead of dismissing it (and the resulting joy in your life), consider lowering your costs and simplify.

This smacks of budgets, and I am not a big fan of budgets. However, I have come to realize in my recent growing up and making choices, that many things become clear if you dig beneath the "I want it because I want it" thinking. For me it comes down to choosing what is important in the longer view. What I mean by that is I may want to get a digital piano. But looking a little further out, I would also like a fun sports car. And a good way to responsibility purchase big items is with a budget that enables you to accumulate some savings.

I don't have a detailed spreadsheet on all my costs, although I was tempted. Instead, I just looked at the major expenses over three months and figured out what I really needed and what could be trimmed. Then I scrutinize my credit card bill each month and see how I did and adjust as necessary for the current month. The hard part is again getting your head straight. Once you have that figured out and don't just impulse buy, the rest is just paying attention.

Sometimes you can make changes in your lifestyle that are lower cost and better for you. When I decided to have a healthier diet, I found that it was difficult to find takeout options that were good for you and started cooking a little more – and my total food bill decreased.

Time – Money Equation

You may have heard the expression "time is money" and I believe it. By spending money wisely, you can get yourself more time to do things you enjoy or are important to you rather than wasting time on mundane tasks.

For example, you can mow your own lawn or pay someone to do it. If you pay $20 to get your lawn cut, you can spend the hour it would take you to do it yourself on another activity. Alternatively, if you have extra time on your hands and help out your neighbor by cutting their lawn, they would pay you $20. It's your choice; for a particular activity, do you want the time or the money?

> **Time you give up = Money saved**
>
> **Money Spent = Time you gain**

Seems simple enough, but sometimes we don't factor the time of routine activities and default to our programming. For example, I used to spend 30 minutes every week going through the massive stack of grocery ads scanning for special deals and coupons. Then easily used another 30 minutes making a list of the specials compiled by grocery stores to visit three or four different stores and search for those specific items to get the best deals. On average I estimate that my coupon savings were about $4.50 (neglecting any added gas).

One day I saw the pile of ads that I had been avoiding and realized that here is my programming in action again – *"Thou shall save every penny – no matter what."* All my grandparents were immigrants from Lithuania so it is understandable that this was their credo. However, this was passed on to their kids and to me. It's very sneaky. The programming is just hiding there from our childhood innocently passing as another truth of the universe and never questioned. Well, now I'm questioning it!

Here's another way to look at my ad search "savings" – would I pay $4.50 for an extra hour in my day? Heck yeah! I would buy a whole carton at that price. So I now gleefully throw all grocery ads in the recycling bin knowing that I have just created myself an extra hour. I don't need to remember to bring my special list, get four cans of tomatoes when I only need one, get the paper towels that don't have the convenient half-sheet perforation or feel bad because I forgot to bring the damn coupons. I now know good deals when I see them so I will make smart decisions at the grocery store (including healthy considerations) without giving up extra time.

The lesson here is to be awake, collect some data and make a decision. If you enjoy going through grocery ads and delight in saving every penny – great! That is your decision and a good choice for you.

A Fresh Start

Most people spend 40 hours of their week at a job to pay the bills. So it's important that you enjoy at least some aspects of your work. If not, figure out why. Is it the department or people around you? Then look within the company for opportunities that are more satisfying for you. There is a whole building of contacts at your disposal; start networking and figuring out where you would like to be and get a plan together on how you can become invaluable to the person running that group.

If there are no viable options internally, start thinking about other companies. Are there workshops or classes you could take to learn new skills that are hot in your industry? Attend or volunteer at relevant conferences to meet new contacts. You may even run into some former associates that could lead you to new opportunities.

If you don't think there are any opportunities that work for you with your current employer, then you may want to explore outside. I've left a company without really understanding why and regretted it. No matter how you feel right now, stop and figure this out. Because if you do not, more than likely you will regret it later and jobs are a challenge to find. Also, doing the work to find the root cause, whatever it may be, will shorten the wait time to happiness. In my case, I tried to run away from my problems with new jobs, new cities, new routines but I eventually found out the problem was with me. So none of that mattered until I fixed me; it was only a matter of time before I would have to deal with the same issue in a slightly different form.

EXERCISE – Not Happy with my Job (30 min)

If you don't like where you are or what you're doing, answer these "why" questions:

Why are you unhappy? _____

_____

_____

Is your job really the issue? _____

Why don't you like your current position? _____

_____

_____

Why don't you like your current employer? _____

_____

_____

What don't you like about your daily routine? _____

_____

_____

Why don't you like your co-workers? _____

_____

_____

What issues do you have with your boss? _____

_____

_____

If you get a clue that the job may not be the issue, go back and re-read the "Obstacles" chapter and examine your answers. Did you miss something important? If that doesn't help, try an honest conversation with a trusted friend or family member. Ask them for their opinion of what may be holding you back. And then just listen. Don't defend, justify or explain – just listen. This will set your confidant at ease so they will reveal everything they have to share. And by focusing on them rather on your responses, you will more likely hear what you need to hear. If there's no one you can trust to get this feedback, consider making an appointment to see a therapist. Many companies have Employee Assistance Programs that include confidential career and mental health professionals; if you need one – make an appointment!

Memos and Meetings

After years of being a single contributor as an Applications Engineer, I was practically delirious when my supervisor offered me a manager position. These were heady days in Silicon Valley (San Jose, California) in the early 90's when Tech was king. It seemed like Mecca for Engineers! I had a team of six people and received a significant bump in salary. I worked hard to keep my tech support team motivated, challenged and happy. Every Friday at MY staff meetings, I would bring in muffins, fruit, milkshakes, whatever they wanted – which I gladly paid for myself in appreciation of their contributions.

The company did 360 surveys, which meant feedback collected from employees, peers and managers. And my results were quite good. (Although I focused on the two "needs improvement" comments out of 48 glowing ones). To pull this off I was putting in long hours at work, planning support phone schedules lying awake in bed or while mowing the lawn on the weekend.

About a year into my new role, it dawned on me that I spent the majority of my job and my waking hours, writing memos and attending meetings. So this was life as a manager?! But I did nothing to change it. It was still cool to be running the show; And I had a jumbo loan for the big house and I was sure things would get better – I just needed to work harder.

Follow your Dream

Ten years later, when I was unhappy with my current job and blamed my boss, I took a leap and set my sights on becoming a movie Director. So I moved to Hollywood and expected to be welcomed with open arms; after all I was an engineer and boy could they use me! Turns out it doesn't quite work that way. I wrote a script – nothing. I applied to the coveted Assistant Director Apprentice program and made it through the grueling preliminary submissions and tests. But during the final in-person interview, I realized that this would not lead to actually being a Director and dropped out.

With my corporate video production experience, I thought that I could become a "Behind the Scenes" videographer and producer. Well, I did that for two feature films but then figured out that it was so easy to do this job that anyone, like the Director's nephew, could easily get the gig – and I wouldn't. So I did a little investigation and found that the role of Script Supervisor would be a great on set job for me.

The Script Supervisor or "Scripty" sits next to the Director during shooting (really cool) and is responsible for the script, tracks what is actually shot, writes the notes for the editor, checks continuity (did the lead actor wear a red or blue tie in the prior scene that was shot 2 weeks ago), works with the camera crew to assign scene numbers and helps the actors with their lines – perfect for an engineer! So I went to Scripty School and after 6 months started begging for any student or ultra-low budget production to establish a resume and after two years made it into the IATSE union. It was a really cool job for 4 years; I worked at all the major studios – wow! But it wasn't helping me become a Director. It was really hard to continuously find work (all crew are contract positions for the duration of the show) and seriously undermined my marriage because of the 12 to 16-hour work days and dramatic salary drop compared to my previous career.

Lessons Learned

In hindsight, I should have been smarter about the Hollywood experience. Talking with people in the industry and asking lots of questions would have helped me to better understand what I was about to face. Then coming up with a plan would have been useful ... before leaving the security of a steady paycheck. Also I discovered that it's dangerous to change your industry and your role and your location. Continuity in some aspects of your old job or community is a wise thing to do. Staying in the same industry would have meant retaining my contacts and buddies. Keeping the same role but in a different industry would allow employers to relate to your skills and your experience; you would just need to learn some new jargon and tweaks on how business is conducted differently in that industry. Don't stack the deck against yourself like I did.

From both of these examples I learnt an important lesson; figure out what you want to do every day then pick a title, industry and job that fits. For example, if you like working with animals and not big on paperwork, explore veterinary technician opportunities.

Oh and don't forget about contacts! So many opportunities have come through connections with people. Friends of friends that know the hiring manager in that coveted company you want to join can be found in the most surprising places. Keep in touch with who you know and meet more people any way you can. You never know who might open the door to your new opportunity.

Making Choices that are right for you

After Hollywood I was not sure what to do next. The new reality of the job market for my old industry had change and I was older. So I decided to give up my aspirations for a high powered career that had been my central identity in favor of a favored city to live and explore other aspects of life that I had been neglecting. The concept of the Life Wheel evolved at this time – balancing different aspects of your life. I let go of my assumptions and expectations of responsibilities and high salary in favor of more flexibility and time – more balance for me.

I did a major introspection analyzing my career history and figured what was common to my string of jobs, what I enjoyed, what I did well. For me, I discovered that technical training was the common thread and the piece of every job that really energized me. Daily activities included: playing with high tech stuff, creating PowerPoint presentations (yes, really), educating and entertaining people while having some control over my schedule.

With that in mind, I crafted a master resume; customizing each job application to be a tailor fit. I figured out what I wanted – really wanted, at the daily activity level. And got it! It was at a significant salary cut but it was doing what I loved and exactly where I wanted to be – near great surfing beaches!

EXERCISE – Analyze your Career (60 min)

1. Complete the following table based on your work history starting with your current job at the top progressing down as far as you can. "Duration" is time you spent at the job. "Joy Level" is the Soul Smiling Quotient that we discussed earlier where 1 is lame and 7 is ecstatic!

Company	Title	Duration (years)	Primary Daily Activities	Salary	Joy Level (1-7)

2. What trends do you see? Start with the joy level. Where were you happiest? _____

3. Circle which daily activities contributed to your higher joy level scores.

4. Was there a common job title that had more of these desirable daily activities? Or is there more of a correlation with the companies and/or industries? _____
_____

5. Craft an imaginary job that has many of the daily activities you enjoy.

   New Job Title: _____

   New Job Description: _____
   _____
   _____
   _____
   _____

Be your Own Boss

Perhaps you've decided that you are done working for someone else and you want to call the shots. Great! I have a little experience with that so here are my thoughts. You will be investing time and probably a good amount of money so do some research to improve your odds of success. Here are some questions to think about:

- Business structure (partnership, corporation, franchise)?
- Product or service?
- What are my strengths and what help do I need?
- Where can I talk to people already doing this?
- How long before I have to start making money?
- How will I reach customers?
- How will I get products made, marketed, sold and supported?
- Who will provide the service and support follow up?
- Is your family behind the idea?

Owning a business is a great deal of work, so pick something you love. Success in a small business is about doing the right things rather than doing things right; and the way to find out is with a Business Plan. Don't skip the business plan. It's a test paper airplane to see if your idea will fly. And if you need to convince partners, banks, investors, a spouse – that's how to do it. Here are the 8 main elements of a business plan you should document and understand:

8 Essential Business Plan Elements

1. Executive Summary – one paragraph that clearly explains what you want to do.
2. Business Description – details of the business, especially what is unique about it.
3. Market Analysis – who needs your product and who is the competition?
4. Management Team – who are the people that will turn this idea into reality?
5. Marketing – how will your customers learn about you and your product/service?
6. Sales Strategy – how will you sell your product & what channels will you use?
7. Funding Needed – how much money do you need & how will you spend it?
8. Financials – how much will you make, at what cost & how soon?

Any investors will want to know up front if your idea pencils out. That is to say, is it a realistic plan and will you make enough money fast enough to entice them to finance your business.

Also consider the common sense tests. For example, if you want to open a restaurant, location is critical. So hang out in the parking lot to see who comes, what times and how many patrons show up. Does that support your projected sales? If you don't get customers to come in the door, nothing else matters.

EXERCISE – Career Goal Review (30 min)

Revisit your Life Wheel and review your "Career" goal. For financial stability, you want to keep your job under control. The best way to do that is have a goal that you monitor regularly. If you are unhappy in your current job as determined in an earlier exercise in this chapter, take some action by crafting a specific goal here. If you are happy at work, then perhaps a goal to prep for a promotion is in order. You could always set a more modest goal of learning a new skill or cultivating better relationships. The choice is yours, so make one.

Career Goal: _____

_____

_____

_____

Engineering a New Life:

Closing Thoughts

Convergence

You may find that your life goals lose their walls. That is to say, one activity flows into the next – convergence! For example, one day I was doing a workout following my routine with my music. The music inspired me to keep moving while resting between sets and found myself practicing West Coast dance swing moves. While in the shower, a solution to a problem that was eluding me just popped into my head which I promptly implemented when I returned to my desk. All these events flowed seamlessly together – which makes for a satisfying day! So the lesson for me was to be open to the moment and not be locked in to the one and only way you have planned in your head. The world is full of opportunities for you - just be awake to see them and then just reach out and take the ones you like!

Time is precious, so whenever possible, combine actions for multiple life goals in the same activity. Let's take the example of a life goal of getting centered once a day. Instead of meditating in your living room, how about a brisk walk to the park, where you can get grounded and perhaps chat with some neighbors on the way back. This way you cover potentially 3 goals: getting centered, connecting with people and some exercise. Another way to describe this is synergy where different aspects of your life come together to form a harmonious flow – more like life.

"To Do" Lists

Some engineering skills should be used with caution. "To Do" lists are one of them. They can be an effective way to deal with stuff you just have to get done. They can also be an addiction – a way to avoid dealing with life and instead get distracted on details. If you feel a rush when you check off an item on your list, this may be you.

One Sunday I was getting ready for a week long business trip and after consulting my "To Do" list, found that I probably had time to do one more. Then I paused, reflected, and said no. I will just leave early for the airport. After all, whatever I don't get done will be waiting for me when I get back. And even if I were to finish my entire list, more things would pop up next week anyways.

To Do List
- ☑ Laundry
- ☑ Pack for trip
- ☑ Wash car
- ☑ Email Jeannie
- ☐ Insurance quotes
- ☐ Water plants
- ☐ Buy new jacket

I had a leisurely drive to the airport, caught the shuttle without stress and had time to spare. While I was in the security line, I notice a passenger had dropped his boarding pass and was scurrying on to the scanners. In a loud voice I caught his attention and directed him to his lost item. He scooped it up gratefully and zipped back to his place in line.

I made my way onto the plane, found my seat and retrieved the items I would need during the flight. A woman stopped nearby and was struggling with getting her carry on in the overhead compartment. She accepted my offer to help and I noticed that her luggage was not going to fit. I confidently suggested she remove an item from the outside pouch to accommodate the space. She complied and I deftly inserted the bag in one smooth motion then smartly clicked the cover in place.

Moments later, a passenger sat in the middle seat next to me. And for some reason I naturally began a conversation that lasted a good portion of the trip and truly made the flight more enjoyable. This is not my normal option as I usual work on my laptop or read science fiction.

While I reflected on the whole experience in a glow and with gratitude, it occurred to me that it all started with putting away my "To Do" List. Instead of getting one more item checked off, I focused on the moment at hand, then the next one. The lesson for me was to leave space in my day to breathe, deal with unplanned problems, notice unexpected opportunities and experience joy.

Although many engineers, executives and financial people will disagree, life – the really important stuff – is not about efficiency. This is where I part from most of my engineer brethren who often treat efficiency as the sacred edict. Finding the best way to get something done will generally save time and/or money but it misses the point. Life is not about the end goal, it's about the process – the process of enjoying, sharing and being grateful for the life we have before us.

Yes, most of us need to work, get stuff done but we also want to lead a joyful life. So figure out where you're spending your time and allocate some of it to relax, meditate and connect with other people without time pressure or feeling that you should always be productive. Don't be at the mercy of others or ghosts in your head. You can decide when work is done and it is time to play.

Less Thinking, more Feeling

In the past I was guilty of way too much thinking, which is not a good thing when you sacrifice feeling. I learned in my marriage that your partner's complaint of a problem doesn't necessarily mean that they want a solution. As an engineer, this is absurd – of course you want to fix the problem. But oh, that is so wrong. If they do not want an answer, your efforts will not be appreciated and you in turn may be resentful when they do not return the expected gratitude. This would be a lose-lose situation which could so easily have been avoided.

Sometimes it's just about being heard, getting validation, sharing and connecting. For this, we need to engage feeling and empathy. And by the way, when you do this honestly and fully with someone – it feels really good. Sometimes solutions are appreciated but find out first if they want one.

Regular Check Ups

You have now created a least 7 life goals that are meaningful to you – congratulations! To make them real it is vital that you review them regularly to see how you're doing. There will be some to admire progress, others that have been side tracked that will need some attention and others may no longer be relevant to you, so you can change those!

EXERCISE – Display Your Goals (30 min)

1. Transfer all the life goals you have created at the end of each of the 7 previous chapters to the master sheet on the next page. There is room for two goals per section.

2. Like the restroom attendant checkbox, date and sign at the bottom of the sheet.

3. Cut out your Life Goals Summary page.

4. Post in a prominent area where you will see your goals daily (e.g. office, master bathroom, kitchen fridge). If other people see your goals, that's great! They will be proud of you, keep you accountable to what you have committed and perhaps even inspire them to make some changes in their own lives. What a hero you are!

5. Review your goals regularly. Sign and date when you assess your progress to renew your commitment. Feel free to modify your goals when appropriate. (There's a fresh copy on the back when your edits expand beyond the page).

Life Goals Summary Sheet 1

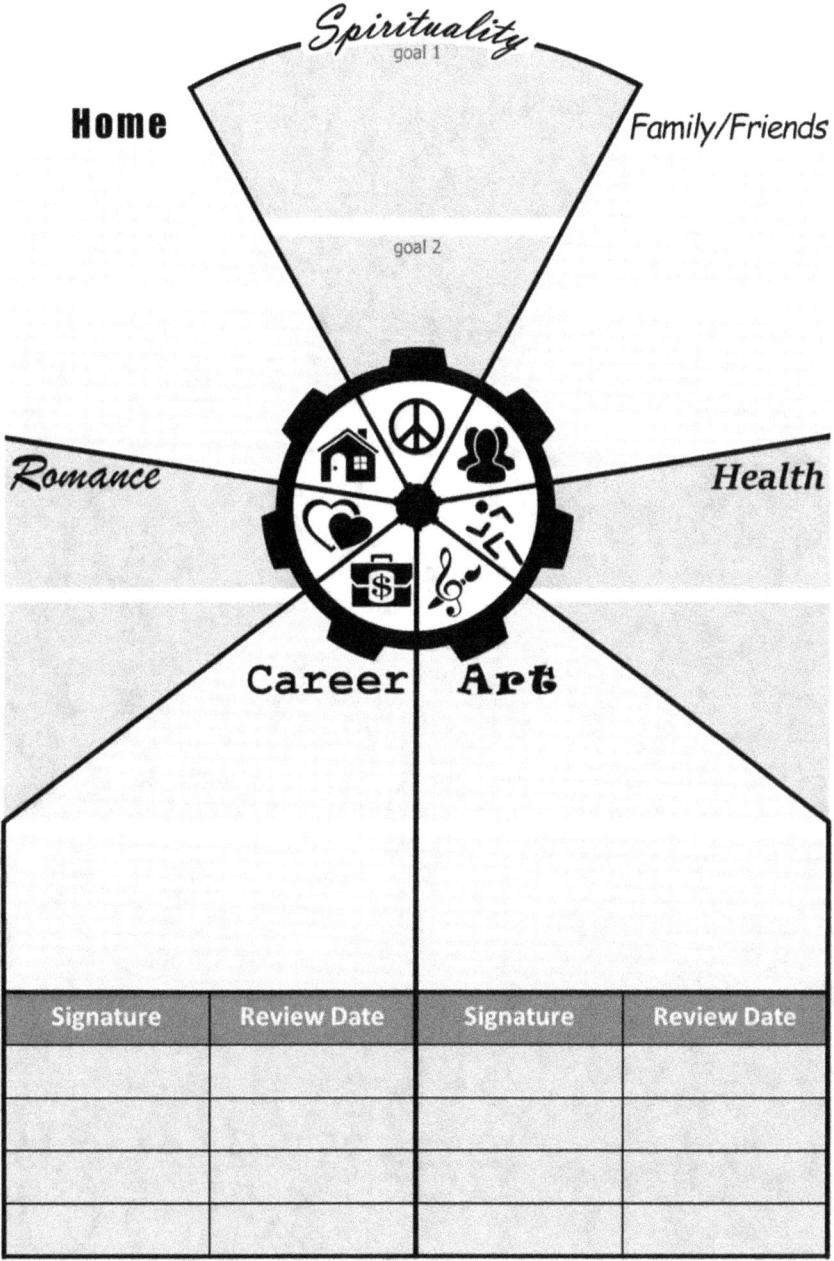

Life Goals Summary Sheet 2

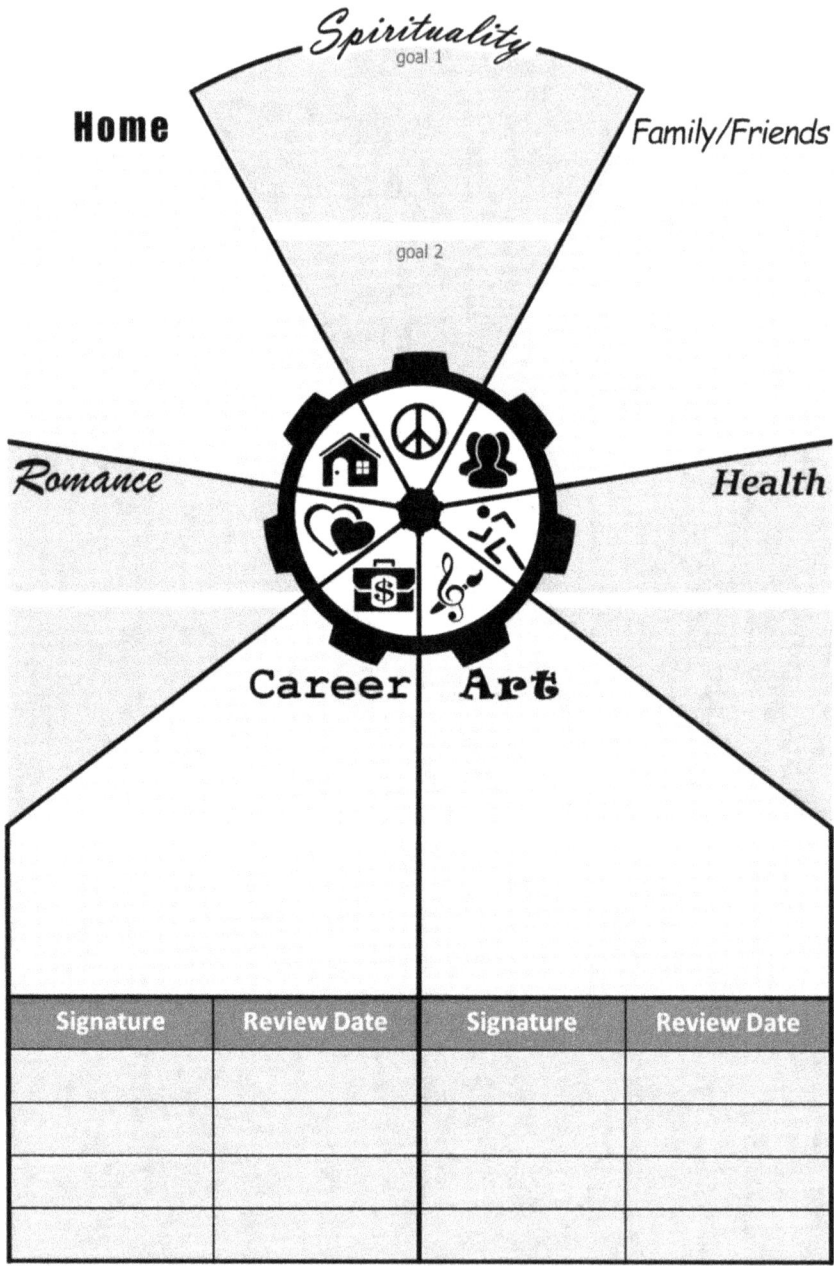

Signature	Review Date	Signature	Review Date

A Passion for Everything

If you engineer a life for you considering your relationships, job, health, activities, you will embrace everything with eagerness and desire – a passion for everything! How cool is that?

And if you're finding one of your goals has dropped to the bottom of the pile after careful consideration, tweak it, or just spend more time on your true passions. You can do absolutely do this. Your life goals are not cast in stone; change them when they are not serving you. At an art class years ago, I was leafing through my sketch book and cringing at some of the pages. The teacher noticed and said: *"If you don't like it, rip it out. You don't need to be reminded of your lesser efforts"*. I gleefully ripped out four pages – and felt so good about it! There's no need for us to torture ourselves with failures. Toss out the negative baggage around you (figuratively and literally) so you can move forward with the passionate stuff!

The secret for me is to fully commit to everything. Engage without hesitation or doubt. Trying a new salmon recipe or a new dance move with gusto produces far more interesting and memorable results – regardless of success! And if you fail, it will be glorious with a great story to share with others. So, really, there is nothing to lose. On the other hand, tepid effort becomes completely unsatisfying, unmemorable and just a waste of time.

Try it with friends by suggesting a cool activity that you have never done – like playing twister. Even if it doesn't happen there will be two or three people who may think you're a fun person for suggesting it. Try it at work by making a bold proposal that no one else would dare speak out loud. If it's accepted, you could have an important project to manage. If not, your boss may still appreciate the out-of-the-box thinking and consider you for that new position he's drafting but hasn't told anyone yet. Try it with your spouse by suggesting a new romantic adventure that ... well I'll let you complete that thought.

Think back to a memorable moment or two in your life. Did you fully commit? I thought so. Do more of that! *"Happiness – being fully present with all activities and thoughts."*[106]

Feedback

Now I could use your help. The experience that is captured in this book was life changing for me. But I have no idea how it will impact you. I would really appreciate some feedback to see if it has made a difference to anyone out there.

> **What impact has this book had on your life?**
> **What did you like best about the book?**
> **Would you like to attend a workshop (specify city if yes)?**

You can email me directly *feedback@engineeringanewlife.com* or provide the info on my website *www.engineeringanewlife.com*

Thank you!

Conclusion

I hope you have found the information presented in this book entertaining and useful in helping you make conscious choices to engineer the life you want. May you find joy at unexpected moments and litter the world around you with happiness!

Answers to the Exercises

You just won the lottery!
Money is a common obstacle and distraction so taking it out of the picture can be a helpful exercise. Also, starting to put your normal thinking aside with your assumptions in this exercise is critical because the point of this book is for you to examine your life from a different perspective. If you did not do the exercise yet, don't be a spectator – be a participant! Go back and do the exercise!

Life Goals
This is your first attempt at capturing what you want. Through the course of the book we will be revising these goals and no doubt you will modify them. The exercise is to get a starting spot for what you want so you should do this quickly. Ideally you will get to some of the core goals from deep in your soul. We can do practical considerations of these goals a little later.

Life Goal Review
I strongly believe that a balanced life is the key to a happy life. If you do not believe this, I urge you to give it a try. It may be uncomfortable to address an aspect of your life that you have not examined carefully before yet it could be the key to discovering new joy. If you do not have at least one goal in each of the seven life wheel categories, go back and add them now.

Category Coloring
Here's a couple of extra Life Wheels if you need them.

Date: _____

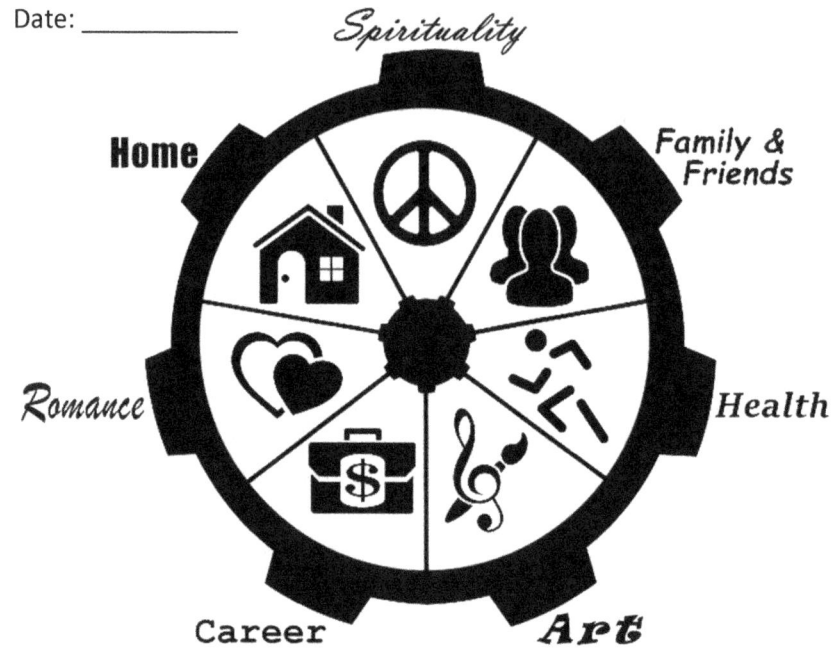

Date: _____

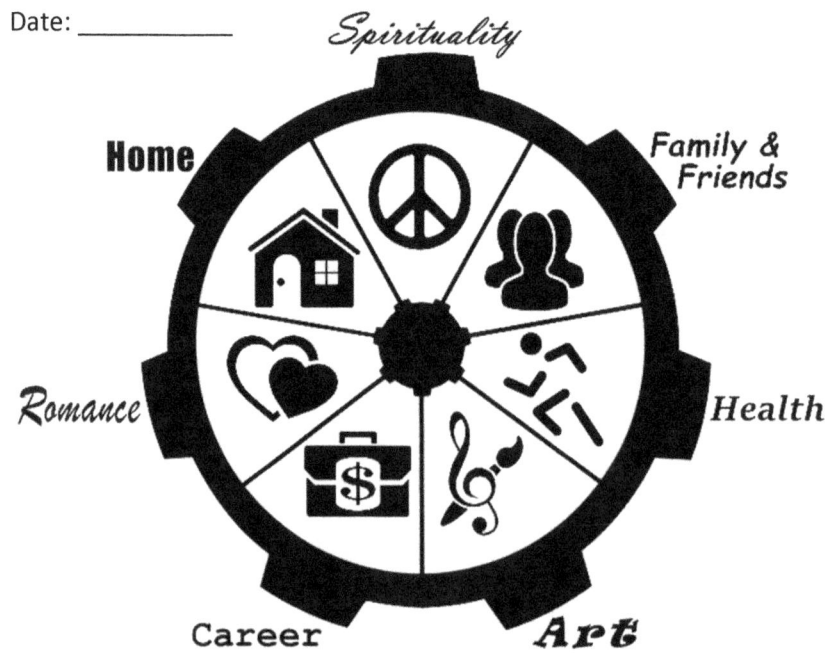

Counting the F's

Most people believe there are 7 letter F's in the identified sentence. There are actually a total of 14. Our brain filters out small words like "an", "if" and "of" as unimportant. Usually this serves us well making more brain power available for other uses. However, this automatic filter can leak into other aspects of our lives and enable us to miss important things around us.

Reflection

Action has a big impact on advancing your intentions. When you write down your commitment to work on your goals and carve out some time out of your busy schedule to figure out what's really going on with you – you've taken the first step. Do you really want a better life? Then sign up for it!

Reflection Now

While you are in the moment and motivated. Do your first reflection now. Not only will it get you started on your long-term goals, you may find that you get some insights on problems pestering your day.

Power of Choice Affirmation

If you did not test out each of the sample affirmations three times out loud in this exercise, go back and try it. As I discuss in more detail in the next chapter, talking out loud engages a different part of our brain and can produce startling results. In reviewing this book I revised countless times and yet when I read a section aloud to someone I suddenly found sentences that had glaring errors even though they had sounded brilliant in my head. And with affirmations in particular, the words have so much more power when spoken – try it!

Time Analysis

Time analysis charts for individual days of the week are provided on the next page if you have irregular days in your week.

Engineering a New Life: Answers to the Exercises

Time	Monday Activities
6AM	
7AM	
8AM	
9AM	
10AM	
11AM	
Noon	
1PM	
2PM	
3PM	
4PM	
5PM	
6PM	
7PM	
8PM	
9PM	
10PM	
11PM	
Midnight	

Time	Tuesday Activities
6AM	
7AM	
8AM	
9AM	
10AM	
11AM	
Noon	
1PM	
2PM	
3PM	
4PM	
5PM	
6PM	
7PM	
8PM	
9PM	
10PM	
11PM	
Midnight	

Time	Wednesday Activities
6AM	
7AM	
8AM	
9AM	
10AM	
11AM	
Noon	
1PM	
2PM	
3PM	
4PM	
5PM	
6PM	
7PM	
8PM	
9PM	
10PM	
11PM	
Midnight	

Time	Thursday Activities
6AM	
7AM	
8AM	
9AM	
10AM	
11AM	
Noon	
1PM	
2PM	
3PM	
4PM	
5PM	
6PM	
7PM	
8PM	
9PM	
10PM	
11PM	
Midnight	

Time	Friday Activities
6AM	
7AM	
8AM	
9AM	
10AM	
11AM	
Noon	
1PM	
2PM	
3PM	
4PM	
5PM	
6PM	
7PM	
8PM	
9PM	
10PM	
11PM	
Midnight	

Time	Saturday Activities
6AM	
7AM	
8AM	
9AM	
10AM	
11AM	
Noon	
1PM	
2PM	
3PM	
4PM	
5PM	
6PM	
7PM	
8PM	
9PM	
10PM	
11PM	
Midnight	

Time	Sunday Activities
6AM	
7AM	
8AM	
9AM	
10AM	
11AM	
Noon	
1PM	
2PM	
3PM	
4PM	
5PM	
6PM	
7PM	
8PM	
9PM	
10PM	
11PM	
Midnight	

Weekly Summary	
Categories	Time/Week
Sleeping	
Working	
Commuting	
Eating	
Hygeniene	
Partner	
Kids	
Friends	
TV/Web/Email	
Total	**168 hrs**

Other Obstacles
It can be difficult to identify what is getting in our way. Imagine you have a magic wand that can eliminate all your roadblocks. At what or who do you need to wave the magic wand to clear your path?

Monster Exorcism
You can't conquer what you can't see. Listing your fears in black and white is the first step to taking control over them. Often when we expose our fears to the light of day, they don't seem quite so scary. The idea of the "Worst Case" column is to deal with the scariest issue; then once that has been tackled, you know that all the rest of the scenarios are manageable.

Confidence
Success begets success. Listing all your talents will encourage you try additional skills and perhaps acknowledge more of the wonderful traits you already possess.

Drawing
I had a few purposes in mind with this exercise. First to show you that you can indeed create art, it is not limited to a few gifted people. Second, I hope that by dividing up the task with a grid gives you a tangible example of how to approach big challenges. Thirdly, by giving you a non-verbal, visual task my objective was to bore your left brain so it would let your creative right side time to be in charge. (More on this in the "Say Hello to your Right Brain" chapter).

Checking Assumptions
You may find that the routines you do are enjoyable and/or necessary. The intent here is to examine what you do and not take things for granted. In looking under some rocks  you may find some gold nuggets that enable you to significantly change your life for the better. Sometimes it's the small things that can make a big difference.

Brainstorming
You may find that by enlisting some friends with this exercise you empower them to tackle some of their own challenges. *"We did this really fun brainstorming thing with Joan last week and solved a tough problem. Maybe she could help us with this situation?"*

Plan for #1 Life Goal
Harnessing the scientific method is another tool to help you with your life plan. It is a good way to tackle goals where there is a central question to answer and you are uncertain how to proceed. It is an objective way of dealing with the problem which should defuse emotional issues and help get to the facts of the matter.

Soul Smiling Quotient
I believe that feeling self-conscious is a sign of issues that are holding us back. To live a whole-hearted life is to throw off these shackles. This exercise is a small step to free our inner child. And it is fun!

Visiting your Right Brain
Like the previous drawing exercise in the "Changing Behaviors" chapter, most people are left brain dominant which is the language, schedule and judging side. We have to give it a boring task in order for it to relinquish control. Creative, abstract activities like drawing are a perfect way to do this. So in addition to have some fun drawing, you also get a break from stress and worry.

First Meditation
If you're doing this right, it should be similar to the previous exercise where you drew your hand. The objective is to give yourself permission to just ... be. No tasks, no schedule, no expectations.

Art Goal Review
I highly encourage one art life goal for you because by its very nature it gives your left brain a break which is the source of all the judgement, pressure and expectations.

Diet/Exercise Choice

I hope that this exercise conveys the idea of conscious choices. You have six different choices of eating/exercise options – pick any one you like. But ... you need to accept the corresponding consequences. And if you don't like the result, pick another option, don't complain about it. The choices are yours to make.

Calorie Cost
Did you have a nice walk? Did you not have a walk AND not have the snack? Either way, I hope you are getting the idea that you choose your

actions that result in predictable results. Your choices allow you to heavily influence your outcomes!

Health Goal Review

Time to pick some health goals. I sincerely hope you have chosen at least one. I have felt the tremendous effect of healthy choices, it ripples through all your other life wheel areas and makes your life much better.

Meeting People

I included this exercise in this section because I have found the road to spiritual awakening requires that you get out of your own head. And one of the most effective ways to do this is to focus on someone else. This exercise should get you thinking about the importance of focusing on others and how it makes you feel.

Spiritual Goal Review

This should be an easy goal to set. There's not a lot to do with a spiritual goal – that's the point. However, it can be challenging to keep it up week after week. If your week is squeezed for time, these types of goals tend to drop off. Resist the temptation because spirituality, in whatever form it takes for you, performs the vital role of re-invigorating your body, mind and soul.

Communication Recall

Here's a tip on this exercise. If you can't recall a recent communication with detail, go have yourself another one then come back to the exercise. In fact, feel free to repeat as many times as you like. Perhaps even combining this with a diary session if you're working on a particular problem.

Golden Rectangle

If you're wondering about the graphic in the section "The Golden Rule", it's the geometric construction of the golden rectangle. It's a ratio, called "phi", that has been known for thousands of years and governs many designs in nature like the famous nautilus shell by recursively inscribing rectangles and drawing a curve between the vertices. The connection to the book is admittedly tenuous although it does illustrate the origin of a beautiful fundamental principle. I love explaining equations! Perhaps the following derivation will lessen someone's fear of math.

The starting point for phi is elegant – internal symmetry. Our goal is to create a picture frame that conforms to the golden rectangle. Let's start with a piece of wood marked with "A" at one end and "B" at the other.

A ──────────────────────────── **B**

Now we need to cut it at point "C" to create the length and width of our golden rectangle frame. The interesting part is how we figure out where we need to cut our piece of wood.

To figure this out, let's call the cut line "C". We will do some virtual comparisons using math to figure out exactly where to cut … snip! Okay, we now have two pieces of wood: AC and CB as shown below. So far so good.

If we stack these two pieces of wood along with another one of the same initial length AB. We stack the three pieces to outline a triangle, in other words that all three pieces are proportional.

$$\frac{AB}{AC} = \frac{AC}{CB} = \Phi$$

That means the ratio of the initial length to the longer cut piece is the same ratio as the long cut piece to the short cut piece. Mathematically, AB divided by AC needs to equal AC divided by CB in order for this to be true. And this ratio is named phi after the Greek capital letter. We can now rearrange the equation and use the fact that AB = AC + CB to simplify:

$$AB * CB = (AC)^2$$
$$(AC + CB) * CB = (AC)^2$$
$$AC * CB + (CB)^2 = (AC)^2$$
$$(AC)^2 - CB * AC - (CB)^2 = 0$$

The form of the equation may look familiar to you. If you think back to high school algebra this is a quadratic equation (because of the squared AC term). And the famous solution for x is shown below[107]. If we replace AC for x, 1 for a, -CB for b and minus CB squared for c, we can solve AC in terms of CB.

$$\text{For } ax^2 + bx + c = 0, \quad x = \frac{-b \pm \sqrt{b^2 - 4ac}}{2a}$$

$$\text{For } x=AC, a=1, b=-CB, c=-(CB)^2, \quad AC = \frac{CB \pm \sqrt{(CB)^2 + 4(CB)^2}}{2}$$

Now it's simply a matter of reducing and cleaning up the equation to see that the ration of AC divided by CB, or phi, is one plus or minus the square root of five all divided by 2. Why the plus or minus? Because it's a quadratic equation, there are two solutions. If we evaluate the root of 5, we find that phi equals either 1.618 or minus 0.618. And if you reciprocate one of these numbers, it will give you the other. In other words, 1/1.618 = 0.618 if you use enough decimal places. Try it!

$$AC = \frac{CB \pm CB\sqrt{5}}{2}$$

$$AC = \left(\frac{1 \pm \sqrt{5}}{2}\right) CB$$

$$\frac{AC}{CB} = \frac{1 \pm \sqrt{5}}{2} = \Phi = 1.618, \text{ or } -0.618$$

$$AC = 1.618 * CB, \text{ or } -0.618 * CB$$

Now armed with this answer, we can build our golden rectangle frame. Let say we want a frame with a height of 10 inches. That means CB is 10" and using our final equation, the length or the frame or AC is 16.2" as shown below – looks like a pleasant proportion. Now if we do the same for the second solution, AC is minus 6.2". The negative sign indicates that the dimension is in the other direction so we have effectively created a second golden rectangle but the equation defines the width rather than the length!

Two Golden Rectangles from 2 Solutions

Get to know someone

I apologize if some of you find this exercise obvious or unnecessary. For me, this exercise would have been useful in the past and perhaps it will help some readers as well. When you have your meeting, make an effort to listen, learn about them and their point of view. You don't have to agree, just hear and acknowledge their opinion.

People Goal Review

If you want more people in your life, more of a support network around you, then you need to set some people goals on your life wheel. You have to put in the effort to establish and maintain relationships. In the past I thought that it was too late for me – I had alienated my family and just not able to make long-term friends. Nope, I was wrong. By putting in the effort and being able to listen to others, I have now good relationships with a small circle of friends and family.

Romance Goal Review

This is the only exercise in the Romance chapter as I didn't feel qualified to dabble in your love life. Let me just say, a little effort in this area can go a long way. Thinking of romantic activities should not be a chore as it was for me. It you find that you can't wait to surprise your love interest with a special gift or Sunday breakfast, you're on the right track. But perhaps a goal would be good to ensure you translate your feelings into action for the most important person in your life. If this does not describe how you feel, figure this out and do something about it. You may want to review the "Obstacles" chapter to help you get some insights.

De-Clutter
This exercise is one I put off for a long time before attempting it. And when I finally did, it made such a difference! I felt so much better being at home. And considering all the time you spend in your house, starting the moment you open your eyes, even some minor changes to improve your nest can have a profound effect on how you feel.

Home Spec
When you start looking for a new home, more than likely no one house will meet all your requirements. So it's important to prioritize your list and know the ones that are "must haves" and any issues that are "show stoppers". Take your time through this process as it will likely be the biggest investment you make.

Home Goal Review
Big or small, you deserve a home goal on your life wheel to get a comfier place to live.

Not happy with my job
You may be tempted to dismiss this exercise because you need to pay bills and you may feel that you have no choice in the matter. Regardless, I would encourage you to do it anyways. Understanding what's going may surprise and this insight may give you some options you had not considered. You spend at least 40 hours a week working, life is so much better if you enjoy what you do.

Analyze your Career
You may find that your subconscious has been guiding your career by leaning each position in a particular direction. When given some latitude on tasks to perform your choices shape what future opportunities are better suited to you. This exercise may reveal this direction so you can commit your full intention towards what you really enjoy.

Career Goal Review
The last two exercises should make your career goal pretty clear. Refine what you have learned and craft a goal that will enable you to pay your bills and be happier at what you do.

Display your Goals
The Life Goal Summary is your certificate of completion for Engineering a New Life – congratulations! Display it proudly, your goals should not be a secret. And when your goals are readily visible where you can review them regularly, you dramatically increase your success.

References

[1] Edgar Dale, *Audiovisual Methods in Teaching*, 3rd edition, Dryden Press 1969.

[2] Pema Chodron, *When things fall apart, heart advice for difficult times*, 1st edition, Shambhala Publications Inc. 2002.

[3] John Koenig, *The Dictionary of Obscure Sorrows*, www.dictionaryofobscuresorrows.com 2014.

[4] Daniel Povinelli, Monique de Veer, Gordon Gallup Jr., Laura Theall, Ruud van den Bos, *An 8-year longitudinal study of mirror self-recognition in chimpanzees*, Neuropsychologia 41, 2003.

[5] Mark Leary, Nicole Buttermore, *The Evolution of the Human Self: Tracing the Natural History of Self-Awareness*, Journal for the Theory of Social Behavior 2003.

[6] V.S. Ramachandran, *A Brief Tour of Human Consciousness*, Pearson Education Inc. 2004.

[7] Allan Pease, Barbara Pease, *The definitive book of body language*, 1st edition, Bantam Books, 2004.

[8] Paul Brocks, *Into the Silent Land: Travels in Neuropsychology*, 1st edition, Atlantic Monthly Press 2004.

[9] Jad Abumrad, Ellen Horne, *Who am I?*, Radiolab.org Podcast Season 1 Episode 1, WNYC Studios February 4th 2005.

[10] Carl Sagan, *Billions and Billions: Thoughts on life and death at the brink of the millennium*, 1st edition, Ballantine Publishing Group 1997.

[11] ACA WSO Inc., *Adult Children of Alcoholic and Dysfunctional Families Fellowship Text*, 1st edition, ACA World Service Organization 2006.

[12] Pema Chodron, *When things fall apart, heart advice for difficult times*, 1st edition, Shambhala Publications Inc. 2002.

[13] William Wynn Westcott, *Numbers: Their occult power and mystic virtues*, VAM Publishing 2017.

[14] John Anthony West, *Serpent in the Sky: The High Wisdom of Ancient Egypt*, Quest Books 1993.

[15] Michael Riordan, Lillian Hoddeson, *Crystal Fire: The Birth of the Information Age*, W.W. Norton & Company 1998.

[16] Nancy Marie Brown, *The Abacus and the Cross: The Story of the Pope Who Brought the Light of Science to the Dark Ages*, 1st edition, Basic Books 2010.

[17] Peter M. Hopp, *Slide Rules: Their History, Models, and Makers*, 1st edition, Astragal Press 1999.

[18] Alexander Jones, *A Portable Cosmos: Revealing the Antikythera Mechanism*, Oxford University Press, 2017.

[19] Bob Brier, Ronald S. Wade, *Surgical Procedures During Ancient Egyptian Mummification*, volume 33, No. 1, Universidad de Tarapaca, Chile 2001.

[20] Andre Parent, *Carpenter's Human Neuroanatomy*, 9th edition, Williams & Wilkins 1996.

[21] Paul Broks, *Into the Silent Land: Travels in Neuropsychology*, 1st edition, Grove Atlantic Inc. 2004.

[22] Jill Bolte Taylor, *My Stroke of Insight: A Brain Scientist's Personal Journey*, Penguin Group 2009.

[23] Michael P. Muehlenbein, *Human Evolutionary Biology*, 1st edition, Cambridge University Press 2010.

[24] The Wachowski Brothers, *The Matrix*, Warner Brothers 1999.

[25] Aldous Huxley, *Brave New World*, Harper Collins Publishing 2006 (first published 1932).

[26] Jad Abumrad, Ellen Horne, *Loops*, Radiolab.org Podcast Season 10 Episode 3, WNYC Studios October 4th 2011.

[27] Jad Abumrad, Ellen Horne, *Falling*, Radiolab.org Podcast Season 8 Episode 3, WNYC Studios September 20th 2010.

[28] Michael Dobbs, *One Minute to Midnight*, Random House 2009.

[29] Jad Abumrad, Ellen Horne, *The Good Show*, Radiolab.org Podcast Season 9 Episode1, WNYC Studios December 14th 2010.

[30] Kate Allgood, *Get Into the Zone: The Essential Guide To High Performance Through Mental Training*, CreateSpace Independent Publishing 2015.

[31] ACA WSO Inc., *Adult Children of Alcoholic and Dysfunctional Families*, meetings.adultchildren.org/meetings, ACA World Service Organization 2018.

[32] University of Florida Department of Wildlife Ecology and Conservation, *United States Fatality Statistics*, Department of Wildlife Ecology & Conservation University of Florida Wildlife – Johnson Lab 2007.

[33] Chantal Kreviazuk, Raine Maida, Kara DioGuardi, *Walk Away (performed by Kelly Clarkson)*, Smelly Songs/EMI April Music Inc. 2004.

[34] Jill Bolte Taylor, *The Neuroanatomical Transformation of the Teenage Brain*, TEDxYouth (Technology, Education & Design), www.ted.com 2013.

[35] Gary Lupyan, Daniel Swingley, *Self-directed speech affects visual search performance*, Department of Psychology, University of Wisconsin-Madison 2011.

[36] Walter B. Cannon M.D. C.B., George Higginson, *Bodily changes in pain, hunger, fear, and rage*, Appleton-Century-Crofts 1929.

[37] Pema Chodron, *When things fall apart, heart advice for difficult times*, 1st edition, Shambhala Publications Inc. 2002.

[38] Media Dynamics Inc., *Ad Exposure and Recall Rates Across Media*, Media Dynamics Inc. 2014.

[39] ACA WSO Inc., *Adult Children of Alcoholic and Dysfunctional Families Fellowship Text*, 1st edition, ACA World Service Organization 2006.

[40] Henry Enfield Roscoe, *John Dalton and the Rise of Modern Chemistry*, Ulan Press 2012.

[41] J.L. Heilbron, *Ernest Rutherford: And the Explosion of Atoms (Oxford Portraits in Science)*, 1st edition, Oxford University Press, 2003.

[42] Walter D. Loveland, David J. Morrissey, Glenn T. Seaborg, *Modern Nuclear Chemistry*, John Wiley & Sons, 2017.

[43] Stephen R. Covey, *The 7 Habits of Highly Effective People*, Running Press Book Publishers, 2000.

[44] John von Neumann, Oskar Morgenstern, *Theory of Games and Economic Behavior*, Princeton University Press 1944.

[45] Rhonda Byrne, *The Secret*, Atria Publishing Group, Beyond Words Publishing 2006.

[46] The Wachowski Brothers, *The Matrix*, Warner Brothers 1999.

[47] B. Buchsbaum, S. Lemire-Rodger, C. Fang, H. Abdi, *The neural basis of vivid memory is patterned on perception*, Journal of Cognitive Science 2012.

[48] P. Pchelin, R. T. Howell, *The hidden cost of value-seeking: People do not accurately forecast the economic benefits of experiential purchases*, The Journal of Positive Psychology 2014.

[49] Pema Chodron, *When things fall apart, heart advice for difficult times*, 1st edition, Shambhala Publications Inc. 2002.

[50] Brene Brown, Ph.D., *The power of vulnerability: Teachings on Authenticity, Connection & Courage*, Sounds True, 2013.

[51] Brene Brown, Ph.D., *The power of vulnerability*, YouTube TED Talk www.ted.com January 3rd 2011.

[52] Scott Heiferman, Matt Meeker, Brendan McGovern, *www.meetup.com* WeWork Companies Inc. 2001.

[53] Frank Thomas, Ollie Johnston, *The Illusion of Life: Disney Animation*, Disney Editions 1995.

[54] Dan Millman, *Way of the Peaceful Warrior: A Book That Changes Lives*, HJ Kramer 2006.

[55] Walter Issacson, *Einstein: His Life and Universe*, Simon and Shuster 2007.

[56] Stephen S. Carey, *A Beginner's Guide to Scientific Method*, 4th edition, Cengage Learning 2011.

[57] Arthur D. Ritchie, *Reflections on the Philosophy of Sir Arthur Eddington*, 1st edition, Cambridge University Press 1948.

[58] Albert A. Michelson, Edward W. Morley, *On the Relative Motion of the Earth and the Luminiferous Ether*, American Journal of Science 1887.

[59] Guillem Anglada-Escude et al, *A terrestrial planet candidate in a temperate orbit around Proxima Centauri*, Nature 2016.

[60] William Bateson, Gregor Mendel, *Mendel's Principles of Heredity*, Courier Corporation 2013.

[61] Gregor Mendel, *Experiments on Plant Hybridization*, Proceedings of the Natural History Society of Brünn 1866.

[62] Arthur Platt, *The Works of Aristotle: De Generatione Animalium*, Oxford Clarendon Press 1910.

[63] Maurice A. Finocchiaro, *The Essential Galileo*, Hackett Publishing Company, Inc. 2008.

[64] James Irwin, *Apollo 15 Hammer-Feather Drop*, YouTube www.youtube.com/watch?v=oYEgdZ3iEKA 1971.

[65] Hugh G. Gauch, *Scientific Method in Practice*, Cambridge University Press 2003.

[66] Jill Bolte Taylor Ph.D., *My stroke of insight: A Brain Scientist's Personal Journey*, Penguin Books 2009.

[67] Marian Annette, *Handedness and brain asymmetry*, 1st edition, Psychology Press 2002.

[68] Deepak Chopra, *The Seven Spiritual Laws of Yoga*, 1st edition, Turner Publishing Company 2010.

[69] Lawrence Robinson, Melinda Smith M.A., Jeanne Segal Ph.D., Jennifer Shubin, *The Benefits of Play for Adults*, HelpGuide.org 2017.

[70] Yuval Noah Harari, *Sapiens: A Brief History of Humankind*, Harper 2015.

[71] J. Verghese M.D., R. B. Lipton M.D., et al, *Leisure Activities and the Risk of Dementia in the Elderly*, New England Journal of Medicine June 19th 2003.

[72] Jordan Daley Fitness Consultant, *www.shapesense.com*, 2015.

[73] Mayo Clinic Staff, *Counting calories: Get back to weight-loss basics*, www.mayoclinic.org/healthy-lifestyle/weight-loss/in-depth/calories/art-20048065, 2017.

[74] CalorieLab, *Calories burned by activity*, calorielab.com/burned, 2018.

[75] Baba Shiv, *George Loewenstein, Antoine Bechara, Hanna Damasio, Antonio R. Damasio, Psychological Investment Behavior and the Negative Side of Emotion*, Vol. 16, Issue 6 Science 2005.

[76] Roberta L. Duyff, *Complete Food & Nutrition Guide*, 5th edition, Houghton Mifflin Harcourt 2017.

[77] Gary Taubes, *The case against sugar*, Anchor 2017.

[78] CalorieLab, *Calories burned by activity*, calorielab.com/burned, CalorieLab Inc. 2018.

[79] Kelli Rae, *42 Cardio Workouts and Other Ideas to Make Exercise Fun and Not Boring*, CreateSpace Independent Publishing Platform 2015.

[80] Patrick Monahan, Jimmy Stafford, et al, *Drops of Jupiter (performed by Train)*, EMI April Music Inc. Blackwood Music and Wunderwood Music 2001.

[81] Dr. Seuss, *How the Grinch Stole Christmas*, Random House Books for Young Readers 1957.

[82] Charles Dickens, *A Christmas Carol*, Chapman and Hall 1843.

[83] Pema Chodron, *When things fall apart, heart advice for difficult times*, 1st edition, Shambhala Publications Inc. 2002.

[84] Brene Brown Ph.D., *The power of vulnerability*, YouTube TED Talk www.ted.com, January 3 2011.

[85] Michael Williams, *Mindfulness: Mindfulness for Beginners - How to Relieve Stress and Anxiety Like a Buddhist Monk and Live in the Present Moment*, CreateSpace Independent Publishing Platform 2016.

[86] Gus Kahn, *Dream a Little Dream of Me*, Gilbert Keyes Music, Words and Music Inc., Don Swan Publications, Tro-Essex Music Inc. 1931.

[87] Bing Crosby, *Bing with a Beat*, RCA Victor 1957.

[88] Jesse Harris, *One Flight Down*, Jesse Harris, Beanly Songs, Sony, ATV Songs LLC 2002.

[89] The Wachowski Brothers, *The Matrix*, Warner Brothers 1999.

[90] Brene Brown, Ph.D., *The Gifts of Imperfection: Let go of who you think you're suppose to be and embrace who you are*, 1st edition Hazelden, 2010.

[91] John Koenig, *The Dictionary of Obscure Sorrows*, www.dictionaryofobscuresorrows.com 2014.

[92] Dan Millman, *Way of the Peaceful Warrior: A Book That Changes Lives*, HJ Kramer, 2006.

[93] Brene Brown, Ph.D., *The power of vulnerability: Teachings on Authenticity, Connection & Courage*, Sounds True, 2013.

[94] Dr. Albert Mehrabian, *Silent Messages: Implicit Communication of Emotions and Attitudes*, Wadsworth Pub Co. 1980.

[95] Kyung W. Chung PhD., Harold M. Chung, Nancy L. Halliday, *Gross Anatomy (Board Review Series)*, 8th edition, Lippincott Williams & Wilkins 2014.

[96] Harry J. Gensler, *Ethics and the Golden Rule,* 1st edition, Routledge 2013.

[97] Philip Van Doren Stern (writer), Frank Capra (director), *It's a Wonderful Life*, Liberty Films 1946.

[98] Leil Lowndes, *How to talk to anyone – 92 little tricks for big success in relationships*, McGraw-Hill Education, 2003.

[99] Jill Bolte Taylor, *The Neuroanatomical Transformation of the Teenage Brain*, TED Ed, ed.ted.com/on/9hy7Fofn 2014.

[100] ACA WSO Inc., *Adult Children of Alcoholic and Dysfunctional Families Fellowship Text*, 1st edition, ACA World Service Organization 2006.

[101] Janet Geringer Woititz, Ed.D., *Struggle for Intimacy*, Health Communications, Inc., 1990.

[102] Frederick Loewe, Alan Jay Lerner, *On the Street where you live*, Chappell and Company Inc. 1956.

[103] Bill Watterson, *The Calvin and Hobbes Tenth Anniversary Book*, Andrews McMeel Publishing 1995.

[104] Ira Glass, *This American Life episode 457: What I did for love*, WBEZ Chicago and National Public Radio, www.thisamericanlife.org/457/transcript 2012.

[105] Ole Brunn, An introduction to Feng Shui, Cambridge University Press 2008.

[106] Pema Chodron, *When things fall apart, heart advice for difficult times*, 1st edition, Shambhala Publications Inc. 2002.

[107] Peter H. Selby, Steve Slavin, *Practical Algebra: A Self-Teaching Guide*, 2nd edition, John Wiley and Sons 1991.

www.ingramcontent.com/pod-product-compliance
Lightning Source LLC
Chambersburg PA
CBHW052024070526
44584CB00016B/1895